# TM 9-252
## WAR DEPARTMENT TECHNICAL MANUAL

**BOFORS
40-MM AUTOMATIC GUN M1 (AA) AND
40-MM ANTIAIRCRAFT GUN CARRIAGES
M2 AND M2A1
TECHNICAL MANUAL**

RESTRICTED DISSEMINATION OF RESTRICTED MATTER — The information contained in restricted documents and the essential characteristics of restricted materiel may be given to any person known to be in the service of the United States and to persons of undoubted loyalty and discretion who are cooperating in Government work, but will not be communicated to the public or to the press except by authorized military public relations agencies. (See also paragraph 18b, AR 380-5, 28 September 1942.)

*by*  *WAR DEPARTMENT*  •  *17 JANUARY 1944*

©2013 Periscope Film LLC
All Rights Reserved
ISBN#978-1-937684-41-9
www.PeriscopeFilm.com

DISCLAIMER:

This manual is sold for historic research purposes only, as an entertainment. It contains obsolete information and is not intended to be used as part of an actual operation or maintenance training program. No book can substitute for proper training by an authorized instructor.

©2013 Periscope Film LLC
All Rights Reserved
ISBN#978-1-937684-41-9
www.PeriscopeFilm.com

*WAR DEPARTMENT TECHNICAL MANUAL*
*TM 9-252*

# 40-mm Automatic Gun M1 (AA) and 40-mm Antiaircraft Gun Carriages M2 and M2A1

*WAR DEPARTMENT* • *17 JANUARY 1944*

---

RESTRICTED *DISSEMINATION OF RESTRICTED MATTER—* The information contained in restricted documents and the essential characteristics of restricted materiel may be given to any person known to be in the service of the United States and to persons of undoubted loyalty and discretion who are cooperating in Government work, but will not be communicated to the public or to the press except by authorized military public relations agencies. (See also paragraph 18b, AR 380-5, 28 September 1942.)

WAR DEPARTMENT
Washington 25, D. C., 17 January 1944

TM 9-252, 40-mm Automatic Gun M1 (AA) and 40-mm Antiaircraft Gun Carriages M2 and M2A1, is published for the information and guidance of all concerned.

[A.G. 300.7 (27 Aug 43)
O.O. 300.7/1106]

BY ORDER OF THE SECRETARY OF WAR:

G. C. MARSHALL,
*Chief of Staff.*

OFFICIAL:
J. A. ULIO,
*Major General,
The Adjutant General.*

DISTRIBUTION: R 9 (4); Bn 9 (2); IBn and H 44 (3); C 9 (8); IC 44 (5).

(For explanation of symbols, see FM 21-6.)

*TM 9-252

# CONTENTS

| | | Paragraphs | Pages |
|---|---|---|---|
| Section I. | Introduction | 1– 5 | 4– 16 |
| II. | Description and functioning of gun | 6– 20 | 17– 73 |
| III. | Description and functioning of carriage | 21– 34 | 73– 98 |
| IV. | Operation | 35– 47 | 98–120 |
| V. | Malfunctions and corrections | 48– 58 | 120–130 |
| VI. | Lubrication | 59– 61 | 130–136 |
| VII. | Care and preservation | 62– 72 | 136–155 |
| VIII. | Inspection and adjustment | 73– 77 | 155–160 |
| IX. | Disassembly and assembly | 78– 89 | 160–187 |
| X. | Sighting and fire control equipment | 90–102 | 187–251 |
| XI. | Ammunition | 103–114 | 252–265 |
| XII. | Organizational spare parts and accessories | 115–117 | 265–277 |
| XIII. | Storage and shipment | 118–120 | 277–282 |
| XIV. | Operation under unusual conditions | 121–125 | 283–290 |
| XV. | References | 126–128 | 291–293 |
| Index | | | 294–301 |

* This Technical Manual supersedes TM 9-252, dated 15 April 1942, including C1, dated 20 October 1942; TB 252-3, dated 13 July 1943; TB 252-4, dated 18 August 1943; TB 252-5, dated 28 October 1943; TC No. 54, dated 1943, section III, in part; and TB 252-6, dated 22 November 1943.

TM 9-252

# 40-MM AUTOMATIC GUN M1 (AA) AND 40-MM ANTIAIRCRAFT GUN CARRIAGES M2 AND M2A1

## Section I
## INTRODUCTION

**1. SCOPE.**

a. This Technical Manual is published for the information of the using arms and services.

b. In addition to descriptions of the 40-mm Automatic Gun M1 (AA) and the 40-mm Antiaircraft Gun Carriages M2 and M2A1, this manual contains technical information required for the identification, operation, inspection, and care of the materiel.

c. Disassembly, assembly, adjustment, and repair of the gun and carriage as may be handled by the using arm personnel are prescribed in this manual. They will be undertaken only under the supervision of an officer or the artillery mechanic.

d. Section X of this manual contains information required for the identification, operation, and care of the items of sighting and fire control equipment which are mounted directly on the gun and carriage. This section also prescribes authorized adjustments and repairs of this equipment. Other items of sighting and fire control equipment, not mounted directly on the carriage, are described in separate Technical Manuals. Refer to section XV for a list of these Technical Manuals.

e. Authorized ammunition for this weapon is described in section XI of this manual, which also provides information required for its identification, use, and care. Refer to section XV for a list of publications containing other pertinent information regarding the ammunition authorized for this weapon.

f. In all cases where the nature of the repair, modification, or adjustment is beyond the scope of the facilities of the unit, the responsible ordnance service should be informed in order that trained personnel, with suitable tools and equipment, may be provided, or proper instructions issued for the performance of the work.

g. This manual differs mainly from TM 9-252, 40-mm Automatic Gun M1 and 40-mm Antiaircraft Gun Carriage M2, dated 15 April 1942, as follows:

(1) Information on the 40-mm Antiaircraft Gun Carriage M2A1 has been added.

(2) New lubrication charts have been added and lubrication instructions revised.

(3) Material on the Director M5 and the Generating Unit M5 has been deleted.

# INTRODUCTION

*Figure 1 — 40-mm Automatic Gun M1 (AA) and 40-mm Antiaircraft Gun Carriage M2A1 — Firing Position — High Elevation*

(4) Information on the Computing Sights M7 and M7A1 has been added.

(5) Instructions on storage and shipment have been added.

(6) More complete instructions for operation under unusual conditions have been added.

## 2. CHARACTERISTICS.

a. The 40-mm Automatic Gun M1 (AA) fires fixed ammunition of armor-piercing and high-explosive types. It fires 1.93 to 2.06 pound shells in rapid bursts at a rate of 120 rounds per minute. The muzzle velocity is from 2,700 to 2,870 feet per second. The maximum effective range, limited by the director, is 3,000 yards.

**TM 9-252**
**2**

### 40-MM AUTOMATIC GUN M1 (AA) AND 40-MM ANTIAIRCRAFT GUN CARRIAGES M2 AND M2A1

Figure 2 — 40-mm Automatic Gun M1 (AA) and 40-mm Antiaircraft Gun Carriage M2 — Firing Position — Low Elevation

TM 9-252
2

INTRODUCTION

Figure 3 — 40-mm Automatic Gun M1 (AA) and 40-mm Antiaircraft Gun Carriage M2 — Travelling Position

## 40-MM AUTOMATIC GUN M1 (AA) AND 40-MM ANTIAIRCRAFT GUN CARRIAGES M2 AND M2A1

b. This gun is intended for duties intermediate between those of the high altitude guns of the 3-inch and 90-mm class and the cal. .50 machine gun. It is effective against dive bombers and low flying aerial targets. It may also be used against ground targets.

c. Most of the shells in the ammunition authorized for this weapon are equipped with tracer. The high-explosive shells are equipped with superquick fuzes which function on impact with a very light material such as an airplane wing. Bursting charges explode the shells unless prior detonation is caused by the functioning of the fuze. The complete rounds weigh from 4.49 to 4.82 pounds. They are loaded into the automatic loader of the weapon in clips of four.

d. The gun may be set on safe, single (semiautomatic) fire, or continuous (full automatic) fire. Cartridges are automatically placed in position for ramming. They are rammed when either of the two firing pedals on the carriage is depressed. The rammed cartridge releases the extractors, allowing the breechblock to close under the tension of the breech ring closing spring. After the breechblock is fully closed, the firing pin is automatically released to fire the piece.

e. The shock of recoil is absorbed by the recoil system which brings the gun to rest after recoil and restores it to battery (counterrecoil). The action of recoil and counterrecoil provides the energy for opening the breech, ejecting the spent cartridge case, and reloading. Empty cartridge cases are ejected from the breech at considerable velocity and are directed toward the front of the weapon by a system of troughing.

f. The weapon may be traversed and elevated manually or by power. Manual control is by two double-handled cranks rotated by operators on seats on the carriage. Power control is by director. Traverse is continuous. The gun may be depressed and elevated from minus 6 degrees to plus 90 degrees. (In power operation, an elevating limit switch actuated by adjustable cam cuts off the power before the gun reaches the upper or lower limits of elevation, permitting it to drift to a stop. The cams are normally set to cut off the power at 0 degree and 85 degrees, respectively.) Spring type equilibrators facilitate elevation and depression of the weapon.

g. The 40-mm Antiaircraft Gun Carriage M2 is of the 2-axle, 4-wheel trailer type. A drawbar with a standard lunette forms the connection between the carriage and the prime mover. Spring suspension is arranged according to the Bofors parallelogram system which permits the wheels to spring independently of each other. The carriage is equipped with electric 4-wheel brakes operated from the prime mover and manually-operated mechanical rear wheel brakes. Combination taillight, stop, and blackout lights are provided.

h. The weapon is normally intended to be fired with the carriage

## INTRODUCTION

Figure 4 — 40-mm Automatic Gun M1 (AA) and 40-mm Antiaircraft Gun Carriage M2A1 — Traveling Position

## 40-MM AUTOMATIC GUN M1 (AA) AND 40-MM ANTIAIRCRAFT GUN CARRIAGES M2 AND M2A1

lowered to the ground with outriggers spread for stability, wheels raised, and chassis stakes emplaced. The axles are equipped with compensating springs to facilitate raising and lowering the carriage. Four screw-operated leveling jacks support the carriage in the lowered position and spirit levels are provided to insure correct leveling. Four chassis stakes are provided to anchor the carriage to the ground in the lowered position.

i. The gun may be fired from the wheels as may be necessary when the weapon is in convoy in combat areas. Under such conditions, the amount of traverse is dependent upon the manner in which the gun is prepared for action. Carriages of late manufacture, and earlier carriages so modified, are equipped with brackets which permit the outriggers to be locked in a traveling position so as to obtain unlimited traverse. The weapon is capable of being towed at high speed on good roads and at medium speed over bad roads and rough terrain.

### 3. DIFFERENCES AMONG MODELS.

a. While a number of minor changes and modifications have been made in the guns and carriages, the majority of these do not affect service of the materiel by the using troops.

b. Gun. Among the differences in the models of the guns which affect service by the using troops are the following:

(1) BARREL ASSEMBLIES. The barrel assemblies on certain of the guns are of Canadian and British manufacture. Their main difference is that the recuperator springs are made of Keystone (rectangular section) wire while these springs in barrel assemblies of United States manufacture are made of round wire. Threads on the parts of Canadian and British manufacture are metric threads; those on parts made in the United States are National Standard threads. Inasmuch as the interrupted threads at the breech end of the barrel, chamber, extractor grooves, rifling form and twist, and exterior and interior diameters are the same on all guns, all barrel assemblies are interchangeable.

(2) BREECHBLOCKS. The breechblock illustrated in this manual (fig. 12) has a removable firing pin bushing screwed into the front face and anchored by a locking pin. Breechblocks of later manufacture have a solid front face. The breechblocks are interchangeable.

(3) RECOIL CYLINDER FLUID. A blend of 60 parts by volume of OIL, hydraulic, and 40 parts by volume of OIL, recoil, light, are now used in the recoil cylinder in place of the glycerine-water mixture formerly used. The changeover to the new mixture is performed only by ordnance maintenance personnel. Cylinders filled with the new mixture are stamped on the front end: "OIL."

TM 9-252

## INTRODUCTION

(4) REGISTER MARK ON GUN TUBE. A lateral machined mark is cut into all gun tubes produced since June 1943. This register mark (fig. 6) is located just to the rear of the flash hider at the bottom of the center line of the tube. It is for convenience in alining the barrel assembly for removal and installation, particularly in the dark.

(5) COOLING SLOTS IN FORE PORTION OF BREECH CASING. Cooling slots are provided in the fore portion of the breech casings of all guns of early manufacture. These slots are eliminated in guns of later manufacture (fig. 9).

c. **Carriages.** Among the differences in the models of the carriages which effect service by the using troops are the following:

(1) 40-MM GUN CARRIAGES M2 AND M2A1. The two carriages have different gear ratios in the hand traversing mechanisms. The gear ratio of the hand traversing mechanism of the 40-mm Gun Carriage M2A1 is approximately three times as high as that of the 40-mm Gun Carriage M2. The rate of traverse per turn of the crank of the M2A1 Carriage is $17\frac{1}{2}$ degrees; of the M2 Carriage, 6 degrees. The principal differences between the hand traversing mechanisms of the two carriages are a larger hand traversing mechanism gear case and long and short traversing mechanism bevel gears of higher ratio on the M2A1 Carriage.

(2) OUTRIGGER EXTENSION BRACKETS. On carriages of late manufacture, and on those so modified, chassis frame outrigger extension brackets with double outrigger hooks (insert, fig. 53) are welded to both sides of the front chassis frame girder. When the outriggers are folded and engaged by the ends of these brackets, continuous traverse of the top carriage is permitted when the weapon is in traveling position. On unmodified carriages of early manufacture, the outriggers, when folded, are engaged by hooks welded directly to the front chassis frame girder; traverse is restricted when the weapon is in traveling position.

(3) OUTRIGGER HINGE COVERS. On carriages of early manufacture, the outrigger hinges were protected by welded steel outrigger hinge covers which were fitted over them when the weapon was placed in traveling position. Canvas outrigger hinge covers are currently being supplied for this purpose.

(4) SAFETY SWITCH AND DUMMY SOCKET. On carriages of earlier manufacture, the electric brake safety switch, together with the dummy socket for the jumper cable, are located on a bracket mounted on the steering drawbar link. On carriages of later manufacture, and on those so modified, the safety switch is mounted on the steering drawbar link but the dummy socket for the jumper cable is mounted on the side of the drawbar (fig. 69).

(5) STRAPS AND CLIPS. On carriages of late manufacture, and on those so modified, web straps and chassis clips are provided to secure the front and rear chassis compensating spring lock handles

## 40-MM AUTOMATIC GUN M1 (AA) AND 40-MM ANTIAIRCRAFT GUN CARRIAGES M2 AND M2A1

in locked position and prevent them from becoming unlocked accidentally.

(6) WHEELS. Some carriages are equipped with wheels of the flat base rim type (fig. 72); others are equipped with wheels of the divided rim type (fig. 73).

(7) TIRES. The standard tire equipment is the 6-ply, heavy-duty, truck-bus type tire with standard heavy-duty tube and flap. Carriages of early manufacture were equipped with 6-ply, heavy-duty, truck-bus type tires and heavy bullet-resisting tubes. Some carriages were equipped with combat tires.

(8) AZIMUTH INDICATORS. Azimuth indicators with two types of faces have been used, the "match the pointer" type (fig. 177) and the "red-black-white (blackout)" type (fig. 178). The majority of the latter have been modified by having their dials painted all black except for alining marks, making them similar to the match the pointer type.

(9) SIGHTS. Weapons are equipped with Sighting System M3 (fig. 161) or the Computing Sight M7 (fig. 162).

## 4. DATA.

**a. General Data Pertaining to the 40-mm Automatic Gun M1 (AA):**

Weight of barrel assembly .................... 295.85 lb
Weight of tipping parts .................... 1,051 lb
Type of breechblock .................... Vertical sliding
Firing mechanism .................... Percussion
Bore:
    Caliber .................... 40-mm or 1.573 in.
    Length .................... 56.24 cal.
    Length .................... 88.58 in.
    Capacity .................... 148.25 cu in.
Chamber:
    Length .................... 12.46 in.
    Taper .................... 0.051 per in.
    Capacity .................... 29.9 cu in.
Rifling:
    Length .................... 75.85 in.
    Number of grooves .................... 16
    Depth of grooves .................... 0.0225 in.
    Width of grooves .................... 0.220 in.
    Twist, nonuniform, increasing from one turn in 45 calibers at the breech to one turn in 30 calibers at the muzzle.
Ammunition:
    Weight of projectile .................... 1.93 to 2.06 lb

TM 9-252
4-5

## INTRODUCTION

| | |
|---|---|
| Weight of propelling charge | 0.65 to 0.72 lb |
| Weight of complete round | 4.49 to 4.82 lb |
| Length of complete round | 17.60 to 17.62 in. |
| Type | High-explosive or armor-piercing |
| Muzzle velocity | 2,700 to 2,870 ft per sec |
| Time of flight at 1,500 yards | 2.0 sec |
| Range (maximum effective, limited by director) | 3,000 yd |
| Type of fire | Single fire or automatic |
| Rate of fire, rapid bursts | 120 rounds per min |
| Capacity of magazine | 7 rounds |
| Capacity of cartridge clips | 4 rounds |
| Maximum recoil | 8.8 in. |
| Normal recoil | 7.4 to 8.3 in. |
| Protrusion of firing pin | 0.099 to 0.114 in. |

b. **General Data Pertaining to the 40-mm Antiaircraft Gun Carriages M2 and M2A1:**

| | |
|---|---|
| Weight of carriage with oil gears | 4,650 lb |
| Weight of complete weapon with accessories | 5,850 lb |

Dimensions:

| | |
|---|---|
| Wheelbase | 126 in. |
| Wheel tread | 55¾ in. |
| Over-all length (traveling) | 18 ft 9½ in. |
| Over-all width (traveling) | 6 ft 0 in. |
| Over-all height (traveling) | 6 ft 7½ in. |
| Over-all height (firing, maximum elevation) | 13 ft 9 in. |
| Minimum turning diameter | 38 ft |
| Road clearance | 14⅛ in. |

Maneuvers:

| | |
|---|---|
| Maximum elevation | 90 deg |
| (Power cut-off by elevation limit switch for director control normally set at 85 degrees) | |
| Maximum depression, top carriage level | −6 deg |
| (Power cut-off by elevation switch for director control normally set at zero degree) | |
| With jacks | −11 deg |
| Amount of traverse | 360 deg continuous |
| Rate of traverse per turn of crank: | |
| M2 Carriage | 6 deg |
| M2A1 Carriage | 17½ deg |
| Rate of elevation per turn of crank | 4 deg |
| Tires | 6.00–20 |
| Tire pressure | 45 lb per sq in. |

## 5. PRECAUTIONS.

a. Before initial loading of the weapon, the outer safety lever should be placed in the "SAFE" position.

13

**TM 9-252**
**5**

## 40-MM AUTOMATIC GUN M1 (AA) AND 40-MM ANTIAIRCRAFT GUN CARRIAGES M2 AND M2A1

 b. For safety, when the breech is open and the gun is not to be fired, always place the hand operating lever in its rear latch bracket where the lever will be in vertical position and the breechblock will be held in open position. With the hand operating lever in this position, the weapon cannot be fired.

 c. The safety straps should be fastened at all times, except when the front and rear chassis compensating spring lock handles are used to unlock the front axles. The axles should never be unlocked unless there are two men to hold the gun stay and two men to hold the drawbar. The locks should be released simultaneously, and the gun stay and the drawbar must be held securely while the locks are being released.

 d. There will be some movement of the carriage, generally in a backward direction, as the carriage is being lowered into firing position. Care should be taken that all members of the crew keep their feet in such positions that there is no possibility that the foot plates of the front and rear leveling jacks will drop on them as the carriage is being lowered.

 e. After the carriage has been lowered, the drawbar and gun stay must be held securely until the chassis compensating spring lock handles have been engaged. Ease the carriage into its lowered position; *do not drop it.*

 f. When the weapon is in firing position, the gun stay handle plungers should always be released so that the gun stay is loose on the axle.

 g. Close and lock the top cover before firing.

 h. Do not exert excessive force on the cartridge remover, for damage to the feed rollers will result.

 i. Under no circumstances will the chassis compensating spring lock handles which lock the front and rear axles be unlocked unless the carriage wheels are in solid contact with the ground. Failure to observe this precaution is likely to result in serious injury to battery personnel.

 j. Do not stand directly in the path of the drawbar or gun stay until the carriage has been securely locked in traveling position. Keep feet from under the foot plates of front and rear leveling jacks.

 k. In traveling, to avoid injury to the personnel, to insure safe road transportation, and to prevent "jackknifing" of the load, the driver should have the load under control at all times by avoiding any slack between the load and the prime mover. On down grades, curves, and rough or slippery roads, the speed should be held to approximately 10 miles per hour.

 l. Lubricate after washing gun and carriage. Do not use high-pres-

## INTRODUCTION

sure washing system for cleaning artillery materiel.

m. Do not allow moisture, dirt, or grit to enter any mechanism or compartment when lubricating, adjusting, inspecting, disassembling, or assembling any part of the weapon.

n. For lubrication, care, and maintenance instructions for use under unusual conditions, refer to section XIV.

o. Do not attempt to remove the barrel assembly unless the breech is open or the breechblock has been removed from the breech ring. To do so will damage the extractors and produce burs in the extractor grooves in the breech end of the barrel.

p. In carrying the barrel assembly, keep it on an even keel. If the muzzle end is raised, there is a possibility that the barrel assembly will become detached from the barrel carrier, permitting the breech of the barrel to strike the ground.

q. Before removing the barrel assembly, be sure that the top cover remains in locked open position and that the breechblock locking pin is in its place in the holes in the breech casing and breech ring. These devices prevent the breech ring and automatic loading tray from moving to the rear after the barrel assembly has been removed.

r. When placing the automatic loader on a flat surface after its removal from the breech casing, great care must be taken to avoid damage of parts. A slight tilting movement at the time of contact may result in such damage.

s. Do not allow any grease or SOLVENT, dry-cleaning, to touch the brake lining or magnet.

t. (Divided rim type wheels) Completely deflate tires before attempting to remove wheel flange retaining nuts. An inflated tire may blow a partially removed flange off the wheel and cause serious injury to personnel.

u. (Flat base rim type wheels) Tap the locking ring during the initial inflation of the tire to seat it firmly in the rim well. Stand aside while inflating the tire to avoid personal injury in case the locking ring is not properly seated in the rim well and flies off the wheel.

v. Disassembly and assembly of Computing Sights M7 and M7A1 by the using arm personnel is permitted only to the extent specifically authorized herein. Turning of screws or other parts not incident to bore sighting, alinement of telescopes, or to the use of the system is expressly forbidden.

w. Keep the Computing Sights M7 and M7A1 clean and in condition for traveling when not in use, the telescopes in their cases and the canvas covers in place.

## TM 9-252
5

### 40-MM AUTOMATIC GUN M1 (AA) AND 40-MM ANTIAIRCRAFT GUN CARRIAGES M2 AND M2A1

x. The following precautions should be taken in the operation and maintenance of the Remote Control System M5:

(1) The elevation limit switch must be "OFF" before engaging or disengaging elevation clutch.

(2) Be particularly careful when orienting in elevation that elevation motor switch at the director is "OFF." Otherwise, a superelevation will be set in, and orienting will be in error by amount of superelevation.

(3) Power should be switched off before cables are connected or disconnected. See that cables are securely held in receptacles before turning power on.

(4) The elevation clutch lever should be put in the "IN" position (top of lever away from coupling) before switching on the power supply.

(5) Never put oil from an unsealed container in oil gears.

(6) Be sure that oil gears have been filled with oil correctly as noted in paragraph 102.

(7) Be sure carriage is level before firing. Correct leveling of carriage cannot be overstressed.

(8) In orienting for elevation or in adjusting oil gears to neutral, unit cover plates should not be removed when dust or rain can get into unit, unless unavoidable. If it is necessary to remove covers under adverse conditions, units should be protected to insure that no dust or rain gets into unit.

(9) Adjustment of backlash in transmitter gearing should be made only by qualified battery electrician.

(10) Do not turn either adjusting screw more than several turns, as otherwise the mechanism may become disengaged.

(11) Exercise care in handling the pilot valve. Do not touch valve proper with hands, as dirt and perspiration are detrimental to it. While valve assembly is out of oil gear, wrap it in a piece of clean paper and keep it in a safe place so as to prevent burring or scratching.

(12) Be sure tube for inserting pilot valve is absolutely clean before using. Clean with SOLVENT, dry-cleaning.

(13) Oil gears must be kept level at all times in order to prevent oil from entering electrical units above oil gear motors.

(14) Do not under any circumstances pull bell housing off motor. If this is done, the rear motor bearings will fall out and replacement of bearings will necessitate use of a special tool which is not issued to using arms. The bell housing is rotated to enable fastening upper portion of oil gear unit to carriage (bell housings which are equipped with two brackets need not be rotated).

## Section II

## DESCRIPTION AND FUNCTIONING OF GUN

### 6. GENERAL.

a. The 40-mm Automatic Gun M1 (fig. 5) consists of the barrel assembly, breech casing assembly, breech ring assembly, breech operating mechanism, a portion of the firing mechanism, automatic loader assembly, automatic loading tray assembly, and recoil mechanism. The gun is supported on trunnions mounted on the sides of the breech casing. The elevating arc is mounted on the under side of the breech casing.

b. The gun barrel fits into the front end of the breech casing and is screwed and locked into the front end of the breech ring. The breech casing forms a chamber for the breech ring, and a support for the barrel assembly, automatic loader, recoil mechanism, and various levers and devices which operate the breech mechanism, firing mechanism, and automatic loader.

c. The automatic loader feeds cartridges into the chamber automatically after the cartridges have been placed in the loader. The recoil mechanism absorbs the backward thrust of the gun in firing and returns it to battery in order that it may be fired again. The recoiling action of the weapon supplies the main source of energy for the operation of the breechblock, automatic loading, and firing mechanisms.

### 7. BARREL ASSEMBLY.

a. The barrel assembly (fig. 6) consists principally of the gun tube, recuperator spring, flash hider, and the necessary mounting devices. The tube is of forged alloy steel. It is rifled with 16 grooves with a nonuniform twist, increasing from one turn in 45 calibers at the breech to one turn in 30 calibers at the muzzle.

b. A flash hider (fig. 6) on the muzzle end of the gun tube protects the operators from temporary blindness from the flash of firing. The flash hider is funnel-shaped; it is screwed against a copper ring and is retained by three set screws.

c. Tubes produced after June 1943, have a register mark at the bottom center line of the tube, just to the rear of the flash hider (insert, fig. 6). This mark is for convenience in alining the barrel assembly for removal and installation, particularly in the dark.

d. The recuperator spring (fig. 6) is located near the breech end of the tube. It absorbs a portion of the energy of the rearward thrust of the gun caused by firing, supplementing the recoil mechanism in absorbing the shock of recoil. It is compressed during recoil. The

**TM 9-252**
**7**

**40-MM AUTOMATIC GUN M1 (AA) AND 40-MM ANTIAIRCRAFT GUN CARRIAGES M2 AND M2A1**

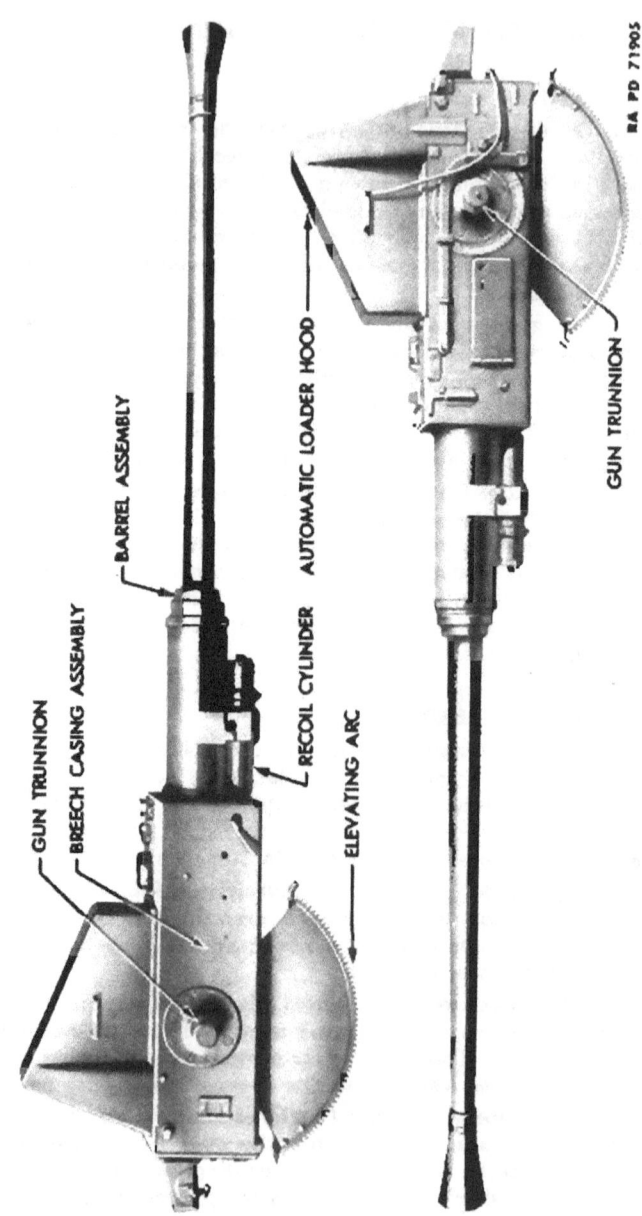

*Figure 5 — 40-mm Automatic Gun M1 — Right and Left Sides*

TM 9-252

## DESCRIPTION AND FUNCTIONING OF GUN

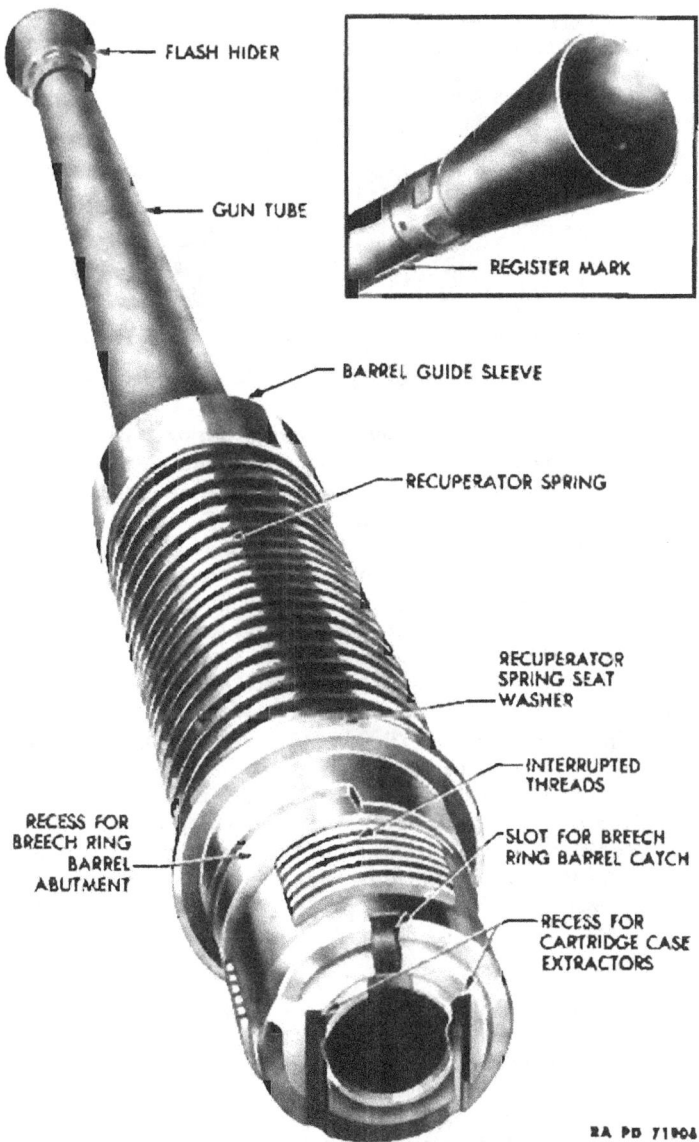

Figure 6 — 40-mm Automatic Gun M1 — Barrel Assembly — Breech End View

TM 9-252
7-8

### 40-MM AUTOMATIC GUN M1 (AA) AND 40-MM ANTIAIRCRAFT GUN CARRIAGES M2 AND M2A1

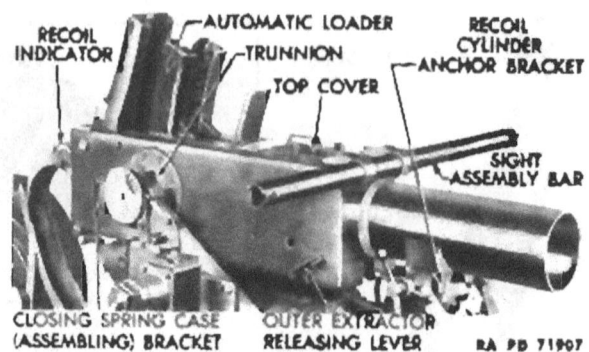

*Figure 7 — Breech Casing Assembly — Right Front View*

energy stored in the recuperator spring during recoil is expended during counterrecoil to force the gun back into battery.

e. The recuperator spring is held in compression against the recuperator spring seat washer and a surface inside the breech casing by the barrel guide sleeve and barrel guide sleeve locking collar. This sleeve centers the tube in the breech casing and provides a surface for the movement of the tube in the breech casing during recoil and counterrecoil. During the removal and replacement of the barrel assembly, the recuperator spring seat washer is retained on the tube by a shoulder or raised surface on the breech end of the tube just forward of the interrupted threads.

f. Interrupted threads (fig. 6) at the rear of the tube are screwed into mating threads in the front end of the breech ring. A vertical slot, cut in the breech end of the tube above the bore, is provided for the breech ring barrel catch. The breech ring barrel catch insures correct assembly and provides a means for locking the barrel assembly in the breech ring. Vertical recesses at each side of the chamber on the breech end of the tube are provided for the cartridge case extractors.

## 8. BREECH CASING.

a. The breech casing (fig. 7) is the housing or supporting unit for the various subassemblies of the gun. The rear portion is rectangular in shape; the front end is tubular. On guns of earlier manufacture, cooling slots were provided in the tubular front end of the breech casing. The casing itself is supported in the carriage mounting by means of flanged trunnions at the sides of the casing. Four hinged or detachable covers give access to the interior of the casing.

b. The recoiling parts of the gun slide in recoil and counterrecoil

TM 9-252
8

## DESCRIPTION AND FUNCTIONING OF GUN

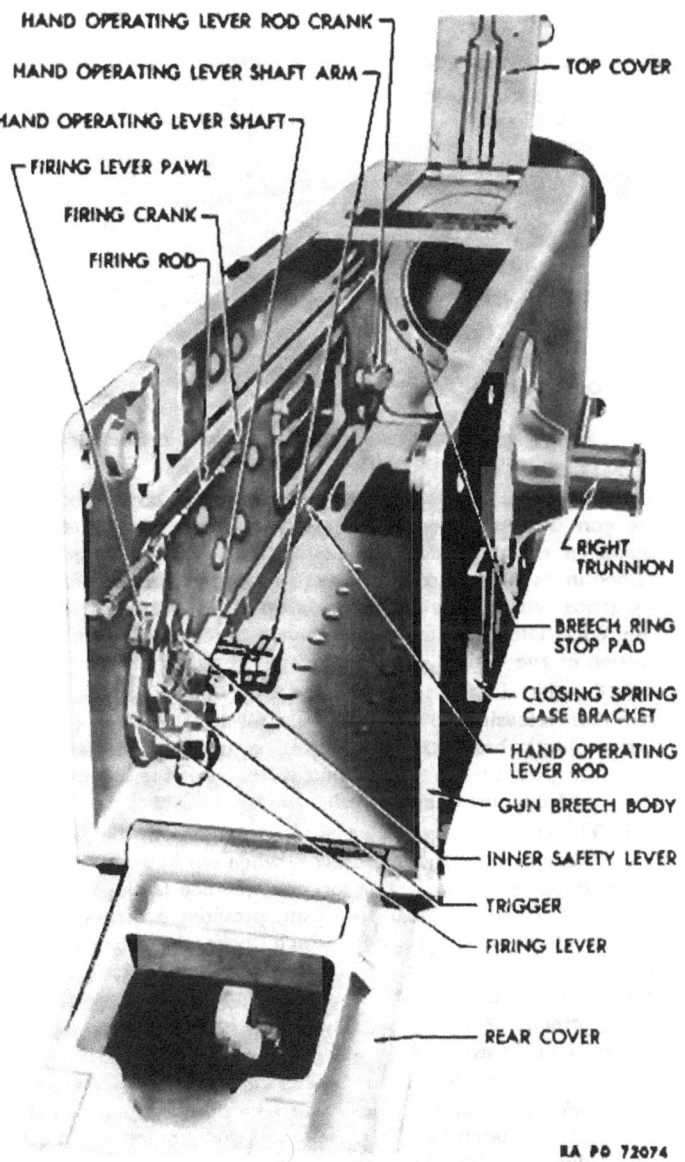

Figure 8 — Breech Casing Assembly — Rear View
— Rear Cover Open

TM 9-252
8

## 40-MM AUTOMATIC GUN M1 (AA) AND 40-MM ANTIAIRCRAFT GUN CARRIAGES M2 AND M2A1

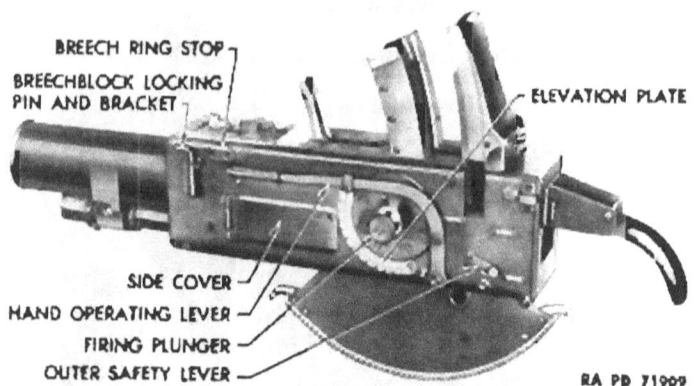

*Figure 9 — Breech Casing Assembly — Left Rear View*

in the breech casing. The recuperator spring on the breech end of the gun is compressed in the tubular front portion of the casing against a surface inside the casing. The breech ring assembly is guided in its backward and forward movement in the casing by bearing strips which slide in channels on the inner sides of the casing (fig. 8). The recoil cylinder anchor bracket, beneath the tubular portion of the casing, supports the recoil cylinder.

c. On the right side of the breech casing (fig. 7) is the outer extractor releasing lever and the breech ring closing spring case (assembling) bracket. The latter is used in disassembling and assembling the breech ring closing spring after the assembly has been removed from the weapon.

d. The automatic loader fits into an opening in the rear of the top of the breech casing and is protected when not in use by the automatic loader hood and shield. The top cover, which is hinged to the upper surface of the casing near the front, provides a means for releasing the breech ring barrel catch which locks the barrel assembly to the breech ring, and also actuates the breech ring barrel catch control arm. Brackets on the front end of the top of the breech casing attach the sight supporting arm to the gun. The sights are described in paragraphs 91 and 92.

e. On the left side of the breech casing (fig. 9) are the breechblock locking pin and bracket, side cover, hand operating lever and front and rear latch brackets, elevation plate, outer safety lever, and firing plunger. The breechblock locking pin is used to lock the breech ring in the breech casing. It is inserted through a slot in the casing and a hole in the breech ring. The side cover has a cammed

22

## DESCRIPTION AND FUNCTIONING OF GUN

inner surface which operates the breech ring outer crank assembly to lower the breechblock automatically during recoil. This cover also permits the removal of the breech ring outer crank assembly and the loading tray attaching bolt.

f. The hand operating lever (fig. 9) opens the breech and prepares the gun for the first round. The outer safety lever prevents the gun from being fired, allows single rounds to be fired, or permits automatic firing. The firing plunger which protrudes through the left trunnion, is the means of contact between the foot pedals on the carriage and the parts of the firing mechanism housed in the breech casing. Cartridge clips are ejected from the automatic loader and breech casing through an opening toward the rear of the left side of the casing.

g. Mounted on the inner left wall and floor of the breech casing (fig. 8) are parts of the hand operating lever mechanism and firing mechanism.

h. The rear of the breech casing is closed by the rear cover (fig. 9) which acts as an abutment for the automatic loader. This cover carries the cartridge case deflector bracket. The recoil indicator is mounted on this bracket. The rear cover is hinged at the bottom and permits removal of the units within the breech casing. This cover has an opening through which the rear end of the loading tray moves in recoil and counterrecoil, and through which the empty cartridge cases are ejected against the cartridge case deflector and down into the cartridge case chute.

i. The hole in the bottom of the breech casing is fitted with the bottom cover which permits removal of the breechblock, breech ring inner cranks, breech closing spring, and extractors without further disassembly of the gun. This cover is not hinged but is retained by a flange at its rear end. The elevating arc is attached to the bottom of the breech casing.

## 9. BREECH RING ASSEMBLY.

a. The breech ring assembly (fig. 10) includes: the breech ring; the breech ring barrel catch which locks barrel assembly to the breech ring; the breechblock with its assembled percussion mechanism parts; the breech ring outer crank assembly which moves the breechblock up and down to open and close the breech; the closing spring assembly which supplies energy to raise the breechblock; the extractor assembly which ejects the empty cartridge cases from the breech and also holds the breech in open position; and the safety plunger which prevents the gun from being fired when the barrel assembly is removed.

b. Breech Ring. The breech ring (fig. 10) is roughly rectangular in shape. It is threaded internally at the front to receive the tube.

**TM 9-252**
**9**

## 40-MM AUTOMATIC GUN M1 (AA) AND 40-MM ANTIAIRCRAFT GUN CARRIAGES M2 AND M2A1

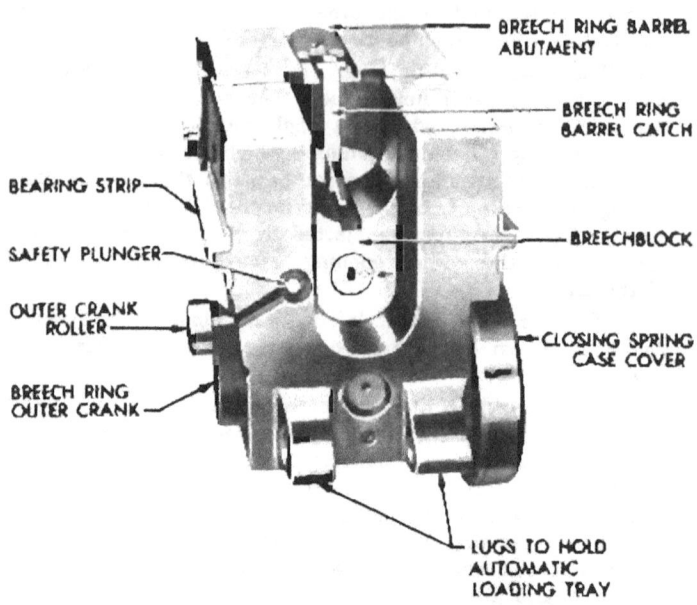

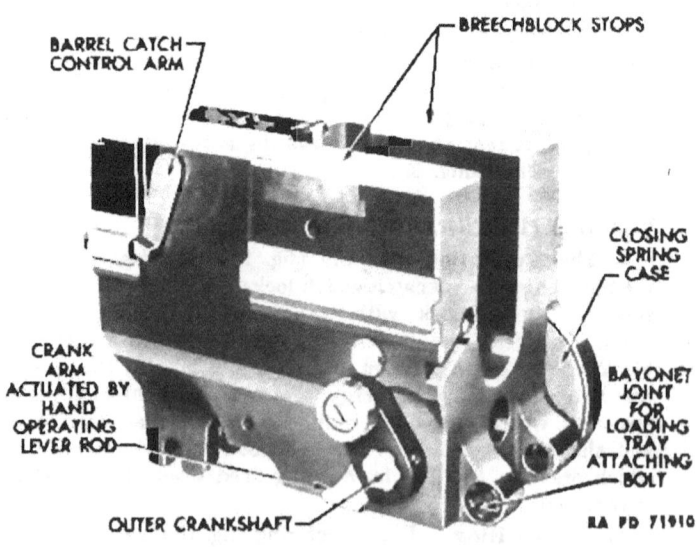

*Figure 10 — Breech Ring Assembly*

## DESCRIPTION AND FUNCTIONING OF GUN

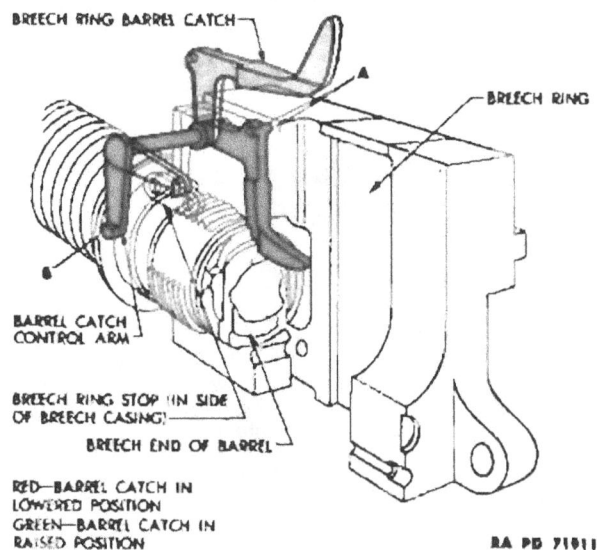

*Figure 11 — Breech Ring Barrel Catch*

It has an opening from top to bottom in the center in which the breechblock slides up and down. Stops, mounted in the upper surface of the breech ring, limit the upward movement when the breechblock closes. The upper rear central portion of the breech ring is cut away in U-shape to provide access for loading and unloading. Two lugs at the rear provide a bayonet joint for the loading tray attaching bolt. Bearing strips on the sides of the breech ring slide in channels on the inside walls of the breech casing when the breech ring is carried backward and forward in recoil and counterrecoil. The extractor spindle arm protrudes from the lower right side of the breech ring.

c. **Breech Ring Barrel Catch.**

(1) The breech ring barrel catch, when lowered, prevents the barrel assembly from rotating to a position where it could be removed from the breech ring. This catch (red in fig. 11) locks the barrel assembly in the breech ring by engaging a slot in the breech end of the tube. The pointed rear end of the catch guides the cartridge into the chamber.

(2) Two lugs (A, fig. 11) at the top rear of the catch ride in a groove in the top cover. When the top cover is opened, this action lifts the catch from its slot in the tube (green in fig. 11) and permits the tube to be rotated and removed from the breech ring.

(3) As the catch is raised, the barrel catch control arm is rotated

TM 9-252
9

## 40-MM AUTOMATIC GUN M1 (AA) AND 40-MM ANTIAIRCRAFT GUN CARRIAGES M2 AND M2A1

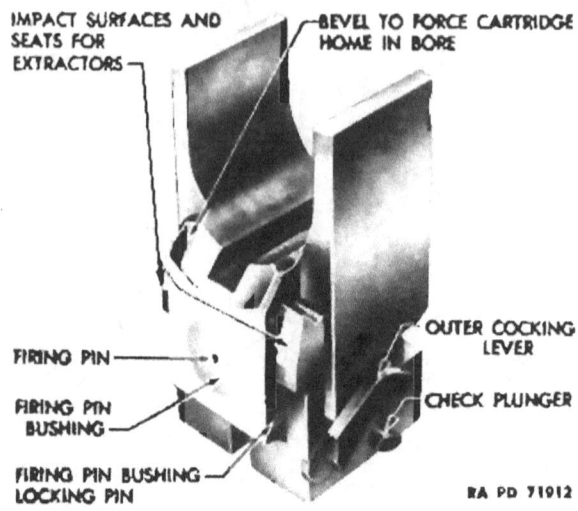

*Figure 12 — Breechblock Assembly — Front View*

to a position where its toe (B, fig. 11) engages the breech ring stop. This locks the breech ring in the casing and prevents it from slipping to the rear when the barrel assembly is removed.

d. **Breechblock Assembly.**

(1) The breechblock (figs. 12 and 13) is roughly rectangular in shape. A U-shaped portion of the upper center is cut away to permit loading and unloading of the weapon when the breechblock is lowered (breech open). A groove is cut in the lower part of the "U" for the finger of the breech ring barrel catch. A square projection on the front surface of the block is beveled at the top to force the cartridge case home as the breechblock slides upward to closed position.

(2) Impact surfaces and hook-shaped seats are provided at the sides of the front of the breechblock for the cartridge case extractors. Horizontal grooves in the sides of the breechblock are provided for the lugs of the breech ring inner cranks which force the breechblock downward and upward to open and closed positions.

(3) The outer cocking lever is located on the lower left side of the breechblock. Its shaft runs from side to side through a bore near the bottom of the breechblock. A groove is cut in the bottom from front to rear for the inner cocking lever which is mounted on the splined shaft of the outer cocking lever. The check plunger occupies another bore which extends from side to side near the bottom of the breechblock.

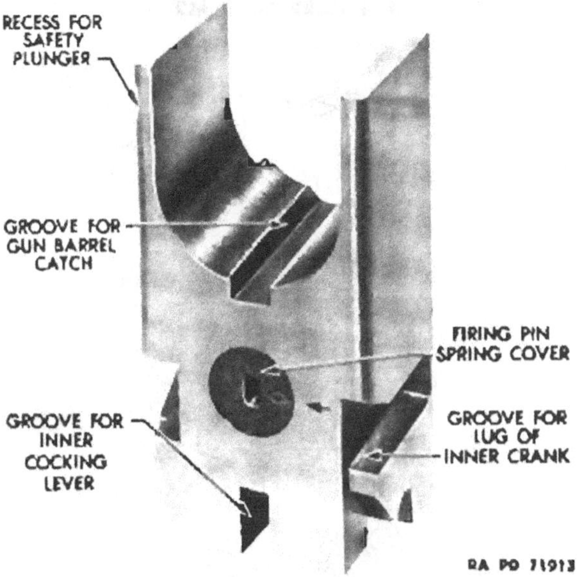

## DESCRIPTION AND FUNCTIONING OF GUN

*Figure 13 — Breechblock Assembly — Rear View*

(4) The breechblock is bored from front to rear. This bore is closed at the front by the firing pin bushing (fig. 12) and at the rear by the firing pin spring cover (fig. 13). This bore houses the firing pin and the firing pin spring. Some breechblocks have a solid front face, no bushing, only a hole for the tip of the firing pin.

(5) The breechblock is recessed on the upper left rear side to receive the breech ring safety plunger (fig. 13). The operation of the safety plunger is described in paragraph 9 g.

e. **Breech Ring Outer Crank Assembly.**

(1) The breech ring outer crank assembly raises and lowers the breechblock to closed and open positions. This mechanism consists of the breech ring outer crank, outer crank roller, outer crankshaft, left inner crank, crankshaft collar, right inner crank, and the closing spring assembly. The latter consists of the closing spring, closing spring case, and closing spring case cover (fig. 14).

(2) When the outer crank is actuated, either by the hand operating lever (hand operation) or the cam surface on the inside of the side operating cover (automatic operation), the crankshaft is rotated. This causes the inner cranks to lower the breechblock by means of lugs which bear on the surfaces of the grooves near the bottom of the breechblock. The rotation of the crankshaft also winds up the closing spring, storing energy which is used to rotate the crankshaft assembly to lift the breechblock and close the breech.

# TM 9-252
## 40-MM AUTOMATIC GUN M1 (AA) AND 40-MM ANTIAIRCRAFT GUN CARRIAGES M2 AND M2A1

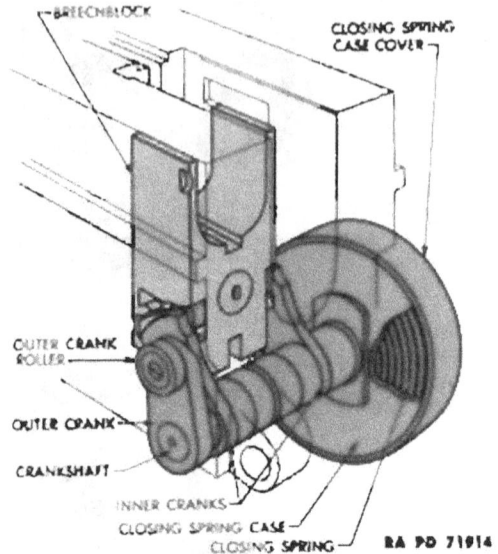

Figure 14 — Closing Spring Raising Breechblock

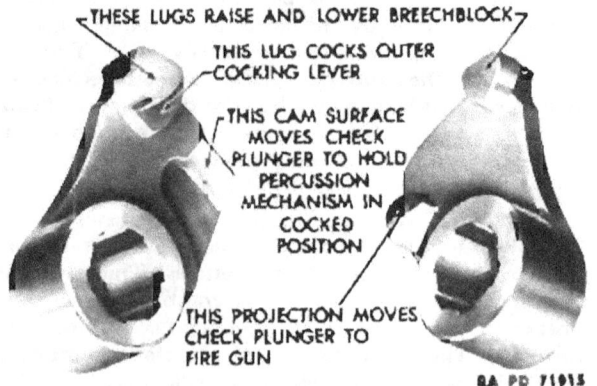

Figure 15 — Left and Right Breech Ring Inner Cranks

(3) The right and left inner cranks (fig. 15) not only bear lugs which raise and lower the breechblock, but another lug on the left inner crank operates the outer cocking lever, and a cam surface on the left inner crank and a projection on the right inner crank move the check plunger (par. 12 f).

28

## DESCRIPTION AND FUNCTIONING OF GUN

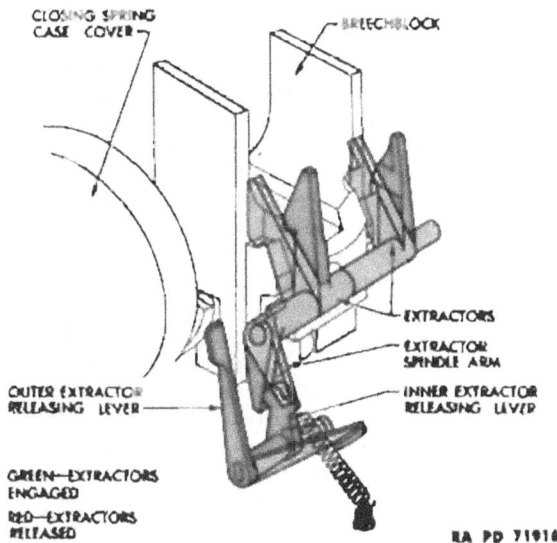

*Figure 16 — Extractor Releasing Lever Assembly*

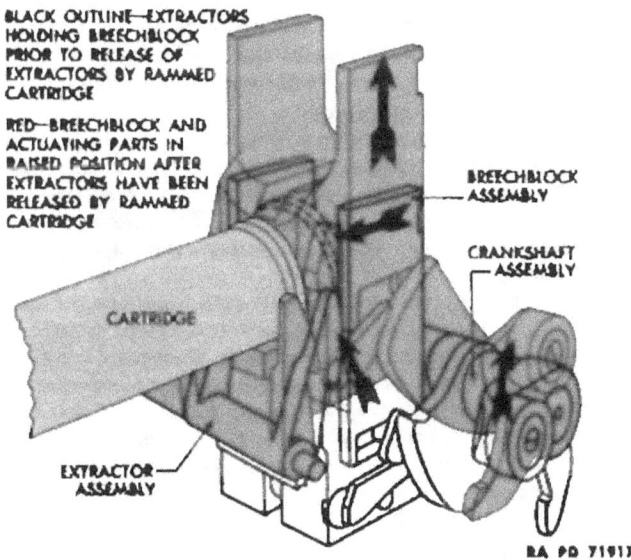

*Figure 17 — Extractors — Released by Cartridge Case*

## 40-MM AUTOMATIC GUN M1 (AA) AND 40-MM ANTIAIRCRAFT GUN CARRIAGES M2 AND M2A1

f. **Extractor Assembly.**

(1) The cartridge case extractor assembly serves two purposes: It holds the breechblock in open position until released by the outer extractor releasing lever (hand operation) or by the insertion of a round into the breech (automatic operation); it releases and ejects the empty cartridge cases from the firing chamber of the gun. The extractor assembly consists of two extractors, extractor spindle, extractor spindle arm, inner and outer extractor releasing levers, spring, and spring stud (fig. 16).

(2) When the breechblock is lowered to open the breech, the extractors are tipped backward until their hooked ends are engaged in seats on the front of the breeckblock (green in fig. 16). In their engaged position, they hold the breechblock in open position. The breechblock is released manually (red in fig. 16) by rotating the outer extractor releasing lever to the rear. The breechblock is released automatically when a cartridge case is rammed into the chamber, the rim of the cartridge case rotating the extractors from their seats on the breechblock as the rim of the cartridge case drives them forward (fig. 17).

(3) The extractors release the cartridge case from the chamber by reversing the procedure illustrated in figure 17. As the extractors are rotated to the rear to engage the breechblock and hold it in open position, their upper lips press on the rim of the cartridge case with such force as to extract the case from the firing chamber and eject it through the breech casing, and out the cartridge chute.

g. **Breech Ring Safety Plunger.**

(1) The rear end of the breech ring safety plunger protrudes from a depression in the left side of the rear of the breech ring (figs. 10 and 18). This device prevents the gun from being fired when the tube is out of its proper position or removed. The safety plunger assembly consists of the safety plunger, compression spring, and safety plunger spring seat.

(2) The safety plunger extends through a bore in the breech ring to contact the breech face of the tube. It is fitted with a tubular seat which fits a depression in the lowered breechblock (A, fig. 18) when the tube is out of its proper position and the compression spring of the safety plunger drives the plunger forward. Thus it locks the breechblock in its down position, putting the breech mechanism out of action. This plunger is pressed backward and its seat is held out of the depression in the breechblock when the barrel assembly is properly assembled in the breech ring (B, fig. 18).

## 10. BREECH OPERATING MECHANISM.

a. The breech operating mechanism consists of the various devices for operating the breechblock which are attached to the breech ring,

TM 9-252
10

## DESCRIPTION AND FUNCTIONING OF GUN

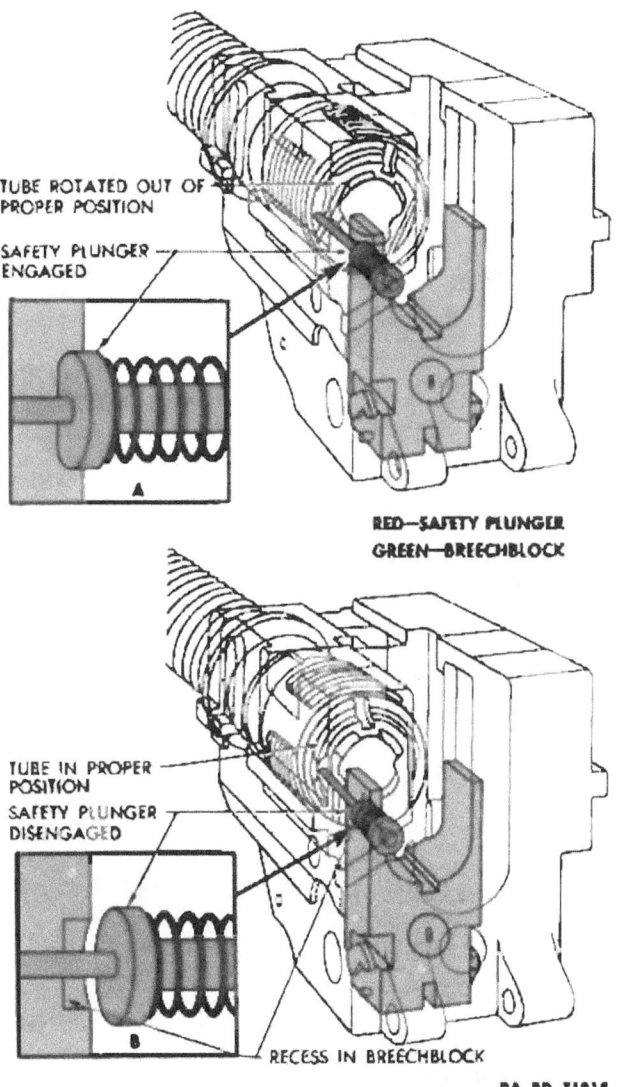

*Figure 18 — Action of Breech Ring Safety Plunger*

**TM 9-252**
**10**

**40-MM AUTOMATIC GUN M1 (AA) AND 40-MM ANTIAIRCRAFT GUN CARRIAGES M2 AND M2A1**

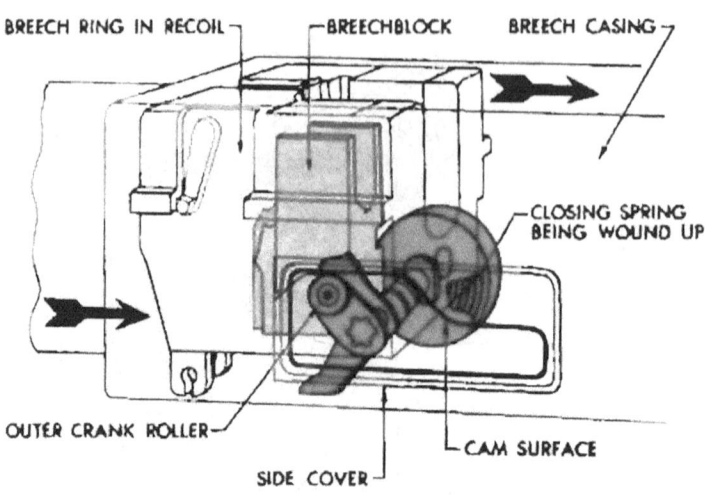

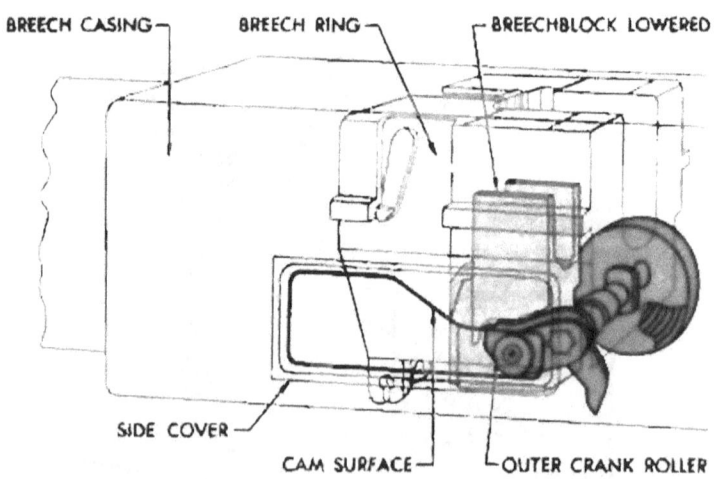

GREEN—BREECHBLOCK
RED—BREECHBLOCK
ACTUATING PARTS

RA PD 71919

*Figure 19 — Cam Action of Outer Crank and Side Cover*

**TM 9-252**

## DESCRIPTION AND FUNCTIONING OF GUN

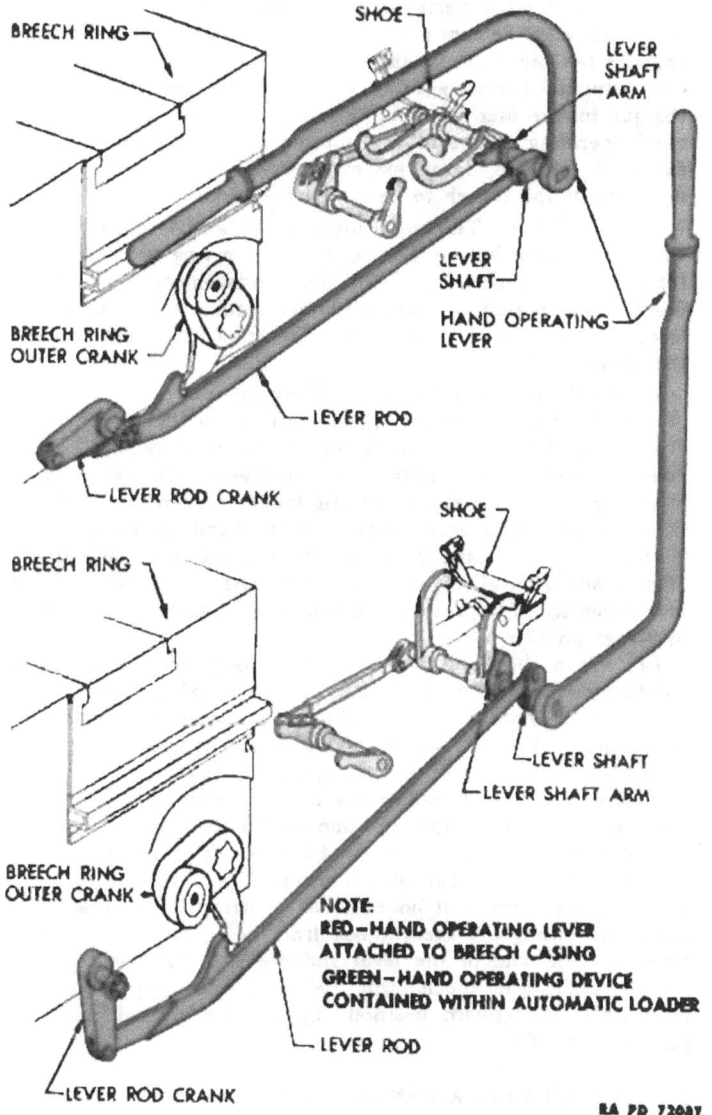

*Figure 20 — Hand Operating Mechanism and Hand Operating Device*

TM 9-252
10-11

**40-MM AUTOMATIC GUN M1 (AA) AND 40-MM ANTIAIRCRAFT GUN CARRIAGES M2 AND M2A1**

together with other parts which are attached to the breech casing. The latter are: the cam surface on the inside of the side cover which engages the outer crank assembly to lower the breechblock automatically; the hand operating lever and mechanism which prepare the gun for the first round by opening the breech and actuating the hand operating device of the automatic loader assembly; and the extractor releasing lever assembly which releases the extractors and permits the breech to be closed manually.

b. Side Cover. The inner side of the side cover has a cam surface which is engaged by the roller of the breech ring outer crank during recoil of the weapon (fig. 19). This engagement causes the outer crank to be rotated, rotating the inner cranks, and causing them to force the breechblock downward. The breech is closed by releasing the extractors (par. 9 f).

c. Hand Operating Lever and Mechanism.

(1) The hand operating mechanism opens the breech manually (red in fig. 20). It also actuates the rammer cocking lever shaft assembly and the feed roller catch release spindle assembly (hand operating device) of the automatic loader (green in fig. 20). The hand operating mechanism consists of the hand operating lever (figs. 9 and 20), lever shaft, lever rod, lever crank, and lever shaft arm (figs. 8 and 20), all of which are mounted on the left wall of the breech casing. The parts of the hand operating device are described in paragraph 17 c.

(2) When the hand operating lever is pulled to the rear, the hand operating lever shaft is rotated. This shaft pulls the front end of the hand operating lever rod downward and to the rear, the motion being limited by the hand operating lever rod crank. A projection on the hand operating lever rod engages the lower arm of the breech ring outer crank and rotates the crank, opening the breech. The breechblock is held in open position by the action of the extractors.

(3) If the gun is not to be fired immediately, the hand operating lever is placed in the rear latch bracket. The projection on the hand operating lever rod will hold the lower arm of the outer crank in place, preventing the breechblock from being raised. At the same time, the rear end of the lever rod prevents the operation of the breech casing firing mechanism. When the hand operating lever is returned to its forward position (fig. 9), the breechblock is freed, permitting the firing of the gun.

**11. BREECH MECHANISM FUNCTIONING.**

a. The breechblock is held in the raised or closed position against the breechblock stops of the breech ring by the tension of the breech ring closing spring. The breech may be opened and closed by hand,

34

## DESCRIPTION AND FUNCTIONING OF GUN

or opened automatically by the action of recoil and closed by the ramming of a cartridge.

b. **Manual Operation.**

(1) To Open Breech.

(a) When the hand operating lever is lifted and pulled to the rear as far as it will go, the lever rotates the hand operating lever shaft. The shaft pulls the front end of the hand operating lever rod downward and to the rear. The movement of the rod is controlled by the hand operating lever crank which pivots in the breech casing. A projection on the rod engages the lower arm of the breech ring outer crank, rotating it in a counterclockwise direction.

(b) The outer crank rotates its shaft, inner cranks, and closing spring case, placing the closing spring under tension. Lugs on the inner cranks bear on the surfaces of the grooves in the sides of the breechblock and force the breechblock downward into the fully open position.

(c) As the breechblock descends, impact surfaces on the front of the breechblock strike the toes of the extractors, rotating them to the rear. If there is a cartridge case in the chamber of the weapon, the lips of the extractors press against its rim, extract it from the firing chamber, and eject it. Hand operation will not always accomplish complete extraction and ejection because of the slowness of the operation. The extractors come to rest directly above the hook-shaped notches in the breechblock into which the hooks of the extractors fit.

(d) When the hand operating lever is moved forward, the outer crank is released and the closing spring tends to assert itself. It raises the breechblock until the hook-shaped notches on the front face of the breechblock are engaged by the hook-shaped projections on the extractors. The breechblock is retained in the open position.

NOTE: The percussion mechanism in the breechblock starts to cock prior to the downward movement of the breechblock. A fully cocked position is reached at the point of initial downward movement of the breechblock (par. 12 g).

(2) To Close Breech.

(a) When the outer extractor releasing lever is pressed rearward, the inner extractor releasing lever is rotated. This lever engages and rotates the extractor spindle arm and extractor spindle, moving the extractors forward. The extractors are withdrawn from the hook-shaped notches on the front face of the breechblock. The breechblock is now free to be raised by the action of the closing spring, provided the hand operating lever is in its forward position.

(b) Under the tension of the closing spring, the inner cranks rotate with the outer crankshaft, raising the breechblock to its closed position against the breechblock stops. After the breechblock is raised, the inner cranks continue to move and their lugs move along the inclines at the rear of the horizontal grooves in the breechblock

## 40-MM AUTOMATIC GUN M1 (AA) AND 40-MM ANTIAIRCRAFT GUN CARRIAGES M2 AND M2A1

until they reach a position where they support the breechblock in such manner that the weight of the breechblock cannot rotate the inner cranks.

c. **Automatic Operation.**

(1) To OPEN BREECH. When the weapon recoils on firing, the roller on the outer crank runs along the cam formed in the inner surface of the side cover. The cam rotates the outer crank and the action is the same as for opening the breech by hand (subpar. b (1), above) except that the extractors hook on their engaging notches on the breechblock as the gun counterrecoils instead of when the hand operating lever is folded forward.

(2) To CLOSE BREECH. The cartridge moving forward by the action of the rammer (par. 16) engages the extractors and rotates them forward, thereby releasing the breechblock. The closing spring asserts itself and the action is the same as for closing the breech by hand.

## 12. FIRING MECHANISM.

a. The gun fires automatically when the breechblock has reached the fully closed position. The percussion mechanism, which actually fires the gun by striking the primer of the cartridge, is housed in the breechblock. The action which fires the piece is the release of the loading rammer which rams the cartridge into the chamber, tripping the extractors, and permitting the breechblock to raise to closed position. Either of two firing pedals on the carriage may be depressed to release the rammer.

b. Means are provided to prevent the gun from being fired, to allow single rounds to be fired, or to permit automatic firing. Means are also provided to stop automatic loading and firing when only one cartridge remains in the feed guides of the automatic loader.

c. The firing mechanism is composed of two principal groups of parts, those attached to the carriage and those attached to or housed in the breech casing. These groups make contact with each other outside the left trunnion of the breech casing.

d. The parts of the firing mechanism attached to the carriage consist of the firing pedals, linkage, and attachments. Their function is to provide means on the carriage for releasing the parts of the firing mechanism housed in the breech casing. The parts of the firing mechanism on the carriage are described in paragraph 29.

e. The parts attached to or housed in the breech casing consist of the percussion mechanism in the breechblock, the rammer catch and check levers of the automatic loader, and the breech casing firing mechanism attached to the inside left wall and floor of the breech casing. The percussion mechanism is automatically cocked and the

## DESCRIPTION AND FUNCTIONING OF GUN

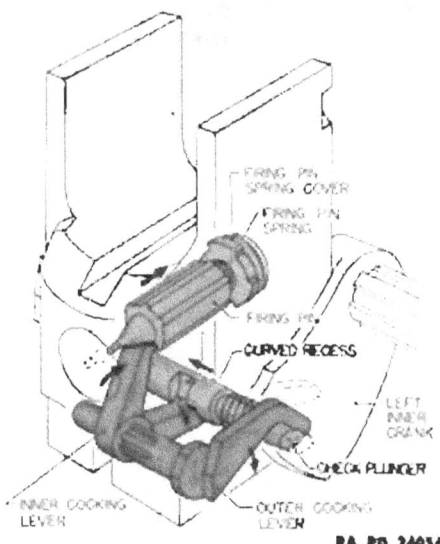

*Figure 21 — Firing Pin — Cocked Position*

rammer is drawn back to firing position when the hand operating lever is pulled all the way backward or when the gun fully recoils and counterrecoils after firing.

*f.* **Percussion Mechanism.**

(1) The percussion mechanism housed in the breechblock consists of the firing pin, firing pin spring, inner cocking lever, outer cocking lever, check plunger, and check plunger spring. The outer cocking lever and the check plunger are actuated by lugs and cammed surfaces on the breech ring inner cranks (fig. 15).

(2) The firing pin has a flat-nosed point and a cup-like rear end (fig. 21). When held in cocked position by the inner cocking lever, the firing pin spring is compressed against the firing pin spring cover at the rear of the firing pin bore in the breechblock. When released, it is driven forward by the spring and its point protrudes through the hole in the center of the firing pin bushing to strike the primer of the cartridge (fig. 22).

(3) The inner cocking lever is an angular piece mounted in a slot in the base of the breechblock and splined on the shaft of the outer cocking lever. One end is ridged; the other is notched; its hub is splined. The outer cocking lever has an angular arm and a splined shaft. The latter causes the inner cocking lever to move with it. The shaft fits into a bore near the front and bottom of the breechblock. The arm of the outer cocking lever rests near the horizontal

37

TM 9-252
12

**40-MM AUTOMATIC GUN M1 (AA) AND 40-MM ANTIAIRCRAFT GUN CARRIAGES M2 AND M2A1**

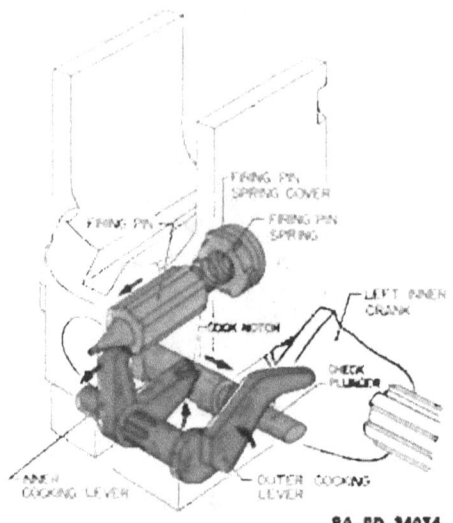

*Figure 22 — Firing Pin — Fired Position*

groove on the left side of the breechblock where it is contacted by a lug on the left breech ring inner crank.

(4) The check plunger is a shaft with two diameters. The end with the smaller diameter has a beveled face; the other end has a cam surface head. The head has flat sides to prevent the plunger from turning in its bore. The check plunger protrudes from both sides of a bore through the lower central portion of the breechblock. The check plunger spring, which fits over the smaller diameter, tends to drive the plunger to the right. A recess is cut in the larger diameter of the plunger. The function of the check plunger is to retain and release the inner cocking lever which retains and releases the firing pin. The plunger is actuated by the breech ring inner cranks.

g. **Percussion Mechanism Functioning.**

(1) To Cock.

(a) When the hand operating lever is pulled backward or the gun recoils, the breech ring outer crank is rotated, rotating the inner cranks and lowering the breechblock. The broader of the stepped lugs on the left inner crank (fig. 15) engages the outer cocking lever, rotating it and the inner cocking lever. The ridge end of the inner cocking lever bears against the shoulder of the firing pin, forcing it to the rear and compressing the firing pin spring (fig. 21). The beveled projection on the right inner crank is moved from contact

## DESCRIPTION AND FUNCTIONING OF GUN

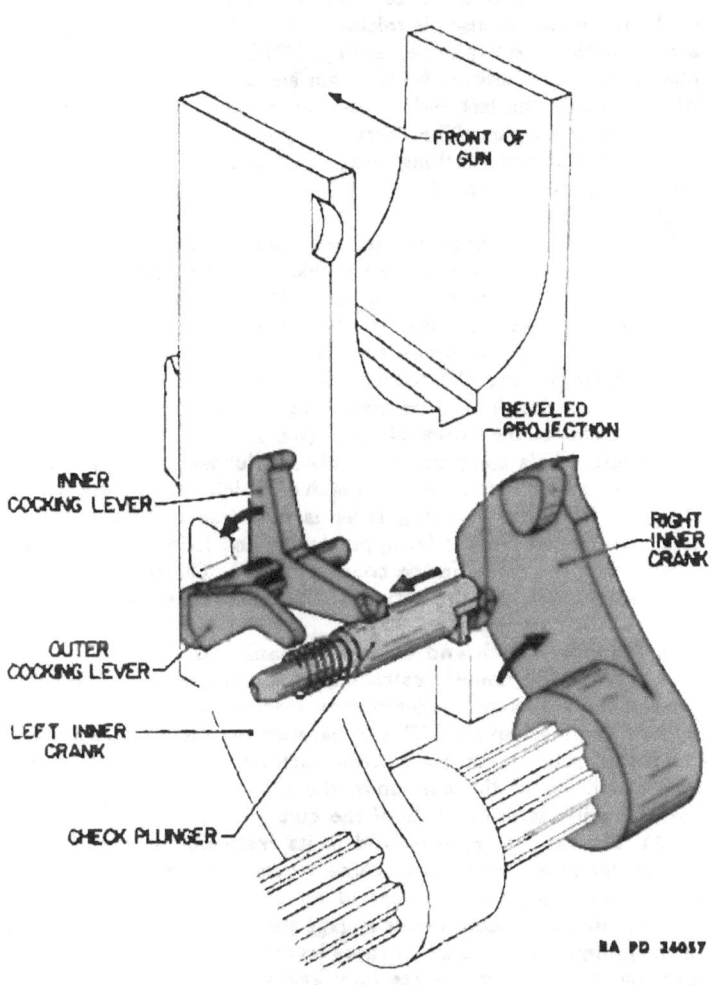

Figure 23 — Check Plunger — Release by Right Inner Crank

## 40-MM AUTOMATIC GUN M1 (AA) AND 40-MM ANTIAIRCRAFT GUN CARRIAGES M2 AND M2A1

with the end of the plunger, freeing the plunger to be forced to the right when released.

(b) When the inner cocking lever has been rotated downward through the recess in the check plunger until its notched end is clear of the recess, the plunger is released and is forced to the right by the action of the check plunger spring. This movement of the check plunger is made positive by the cam surface on the left inner crank which engages the left end of the plunger and forces and holds the plunger to the right. The check plunger engages the cock notch in the end of the inner cocking lever, retaining the lever and firing pin in cocked position (fig. 21).

(2) To FIRE.

(a) When a cartridge is rammed into the breech, releasing the extractors and causing the inner cranks to rotate under the action of the breech ring closing spring, these cranks lift the breechblock. The left end of the check plunger is released from the cam surface of the left inner crank as this crank rotates.

(b) After the breechblock arrives in closed position, the beveled projection on the right inner crank engages the cam surface head at the right end of the check plunger (fig. 23) and forces the plunger to the left. This compresses the check plunger spring and releases the inner cocking lever, the cock notch of which was held by the check plunger. The inner cocking lever is now free to rotate under the action of the compressed firing pin spring and its notched end swings upward into the recess in the check plunger. The released firing pin is carried forward by the action of the compressed firing pin spring and fires the cartridge (fig. 22).

h. **Rammer Catch and Check Mechanisms.**

(1) While the rammer catch and check mechanisms (fig. 38) are parts of the automatic loader control mechanism, they also are parts of the firing mechanism. This is because the release of the rammer and the consequent ramming of the cartridge, releasing of the extractors, and firing of the gun upon the closing of the breech are all operations in the firing cycle of the gun.

(2) The rammer shoe is held in its rearward or cocked position by the upthrust rear ends of three levers mounted on the rammer catch lever axis pin in the rear of the automatic loader base. The raised rearends of these levers engage the shoulder on the under side of the rammer shoe. Each of these levers is the last link of a holding mechanism. All three levers may engage the rammer shoe at one time; or any one or two of them may prevent the rammer from moving forward. All must be depressed before the rammer shoe is released.

(3) These mechanisms are: the automatic catch and release mechanism (par. 17 d) which insures that the rammer cannot be

TM 9-252
12

## DESCRIPTION AND FUNCTIONING OF GUN

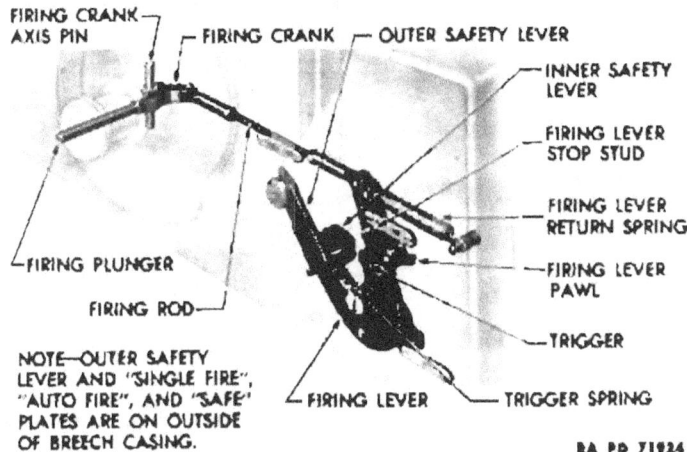

*Figure 24 — Breech Casing Firing Mechanism — Phantom View*

released until the gun reaches the end of conterrecoil; the feed control check and release mechanism (par. 17 e) which can be set to stop automatic loading and firing when only one cartridge remains in the feed guides of the loader; and the firing mechanism check and release mechanism (par. 17 f) which is controlled by the safety lever.

i. **Breech Casing Firing Mechanism.**

(1) The breech casing firing mechanism (fig. 24) consists of the parts of the firing mechanism which are mounted on the left wall and floor of the breech casing (fig. 24). It includes the safety lever and its mechanism. The breech casing firing mechanism converts the movement of the firing plunger in the left gun trunnion into movement which releases the rammer and fires the gun. The firing plunger is actuated by depressing either of the firing pedals on the carriage.

(2) The firing lever (fig. 24) is roughly U-shaped. It is pivoted on a bracket fastened to the floor of the breech casing. Its longer upper arm is pulled forward by the firing rod and crank when the firing plunger is pressed inwardly by the action of depressing either of the firing pedals. It is returned to its normal position by the firing lever return spring.

(3) The firing lever pawl (fig. 24) is pivoted between lugs on the rear of the longer arm of the firing lever. The trigger is located in the "U" of the firing lever and pivoted on a pin in the left wall of the breech casing. When the firing lever is pulled forward, the firing lever pawl contacts the trigger and rotates the trigger on its axis pin. The trigger rotates the rammer control spindle trigger arm of the automatic loader to release the rammer shoe and fire the gun.

41

TM 9-252
12

**40-MM AUTOMATIC GUN M1 (AA) AND 40-MM ANTIAIRCRAFT GUN CARRIAGES M2 AND M2A1**

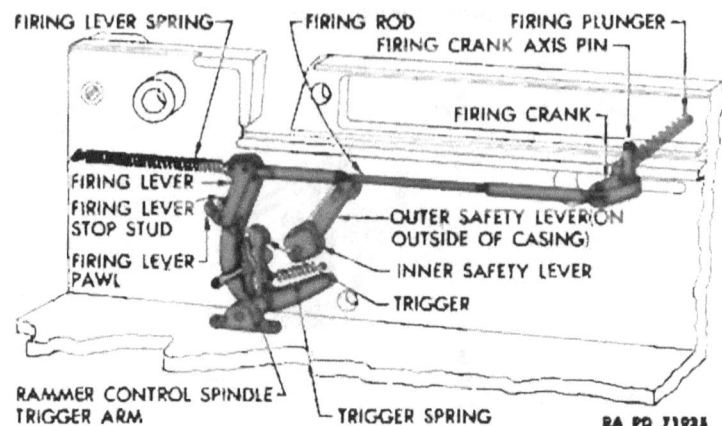

Figure 25 — Firing Mechanism — "Single Fire" — Prior to Firing

When rotation has stopped, the trigger is returned to its normal position against its stop on the firing lever shaft by the trigger spring and the left rammer check lever plunger spring.

(4) The firing lever pawl is tripped by the firing lever stop stud to release the trigger and produce "single fire" action. The rotation of the trigger is controlled by the position of the inner safety lever.

(5) The outer safety lever (fig. 24) is mounted toward the rear on the outer left side of the breech casing. Its head or handle has a spring-loaded plunger which permits it to be locked in holes in either of three positions. Movement of the lever rotates the cam-faced inner safety lever, the surfaces of which control the action of the trigger. In the rearmost or "SAFE" position of the outer safety lever, the gun is prevented from being fired. The central or "AUTO FIRE" position of the lever permits automatic firing. The foremost or "SINGLE FIRE" position of the lever allows single rounds to be fired.

(6) SAFE. With the outer safety lever in "SAFE" or rearward position, the inner safety lever is turned so that its cammed surface restricts the rotation of the trigger to such an extent that the trigger cannot move sufficiently to rotate the rammer control spindle trigger arm and release the rammer.

(7) SINGLE FIRE. With the outer safety lever in "SINGLE FIRE" or forward position, the inner safety lever is rotated until it is completely out of contact with the trigger (fig. 25). When the firing lever is pulled forward by the firing rod, the firing lever pawl rotates the

42

TM 9-252
12

## DESCRIPTION AND FUNCTIONING OF GUN

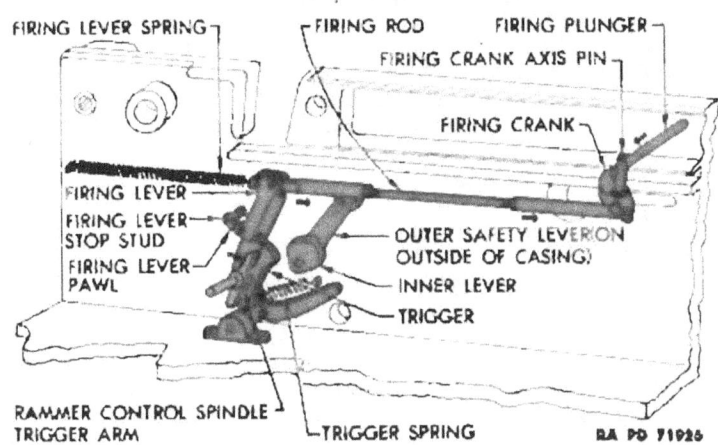

*Figure 26 — Firing Mechanism — "Single Fire" — Firing*

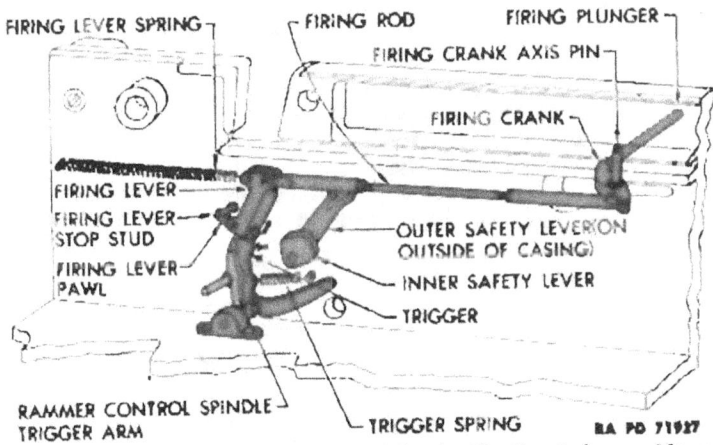

*Figure 27 — Firing Mechanism — "Single Fire" — Release After Single Fire*

trigger to fire one round (fig. 26). This is done by rotating the rammer control spindle trigger arm which releases the left rammer check lever from the rammer shoe. Then the firing lever pawl is tripped upon contacting the firing lever stop stud and the trigger is released and snapped back to its original position by the trigger spring (fig. 27).

(8) AUTOMATIC FIRE. With the outer safety lever in "AUTO FIRE" or central position, the inner safety lever is rotated sufficiently

43

TM 9-252
12-14

**40-MM AUTOMATIC GUN M1 (AA) AND 40-MM ANTIAIRCRAFT
GUN CARRIAGES M2 AND M2A1**

to prevent the trigger from being rotated enough to be released by the firing lever stop stud. It does, however, permit the trigger to rotate far enough to rotate and hold the rammer control spindle trigger arm and depress the left rammer check lever. Uninterrupted firing will continue until the firing pedal is released or the supply of ammunition in the automatic loader is insufficient to operate the gun.

## 13. AUTOMATIC LOADER.

a. The automatic loader (figs. 28 and 29) is both a cartridge magazine and a loading device. Cartridges, in clips of four, are inserted in the top of the loader. They are fed singly to the automatic loading tray, the clips being removed automatically in the process. The cartridges are rammed into the chamber of the gun, tripping the extractors and permitting the breechblock to be raised after which the gun fires automatically.

b. Cartridges may be loaded manually or automatically. After the mechanism has been manually prepared for automatic loading, the feed mechanism of the loader feeds the loading tray with a continuous supply of cartridges. Provisions are made to insure that only one cartridge is fed onto the loading tray at a time, that automatic loading and firing can be stopped when only one cartridge remains in the feed guides and one cartridge on the loading tray (to eliminate the necessity of manual reloading), that the rammer cannot be released until the gun reaches the end of counterrecoil, and that the rammer can be latched out of action to prevent accidental discharge of the weapon.

c. The front and rear cartridge guides and the upper part of the feed mechanism protrude from the top of the breech casing and are covered by the automatic loader hood when the gun is not in action. All other parts of the loader are contained within the breech casing (fig. 5).

d. All parts of the loader with the exception of the loading tray and rammer remain in a fixed position in the breech casing. The loading tray is bolted to the rear end of the breech ring and recoils and counterrecoils with it. The motion of the loading tray is the main source of energy for the operation of the feed mechanism. The rammer moves independently of the loading tray to ram the cartridges.

e. The principal mechanisms of the automatic loader are the feed mechanism, automatic loading tray, cartridge rammer assembly, and automatic loader control mechanism.

## 14. AUTOMATIC LOADER FEED MECHANISM.

a. The feed mechanism (figs. 30 and 31) comprises those parts of the automatic loader which feed the loading tray with a continuous

TM 9-252
14

## DESCRIPTION AND FUNCTIONING OF GUN

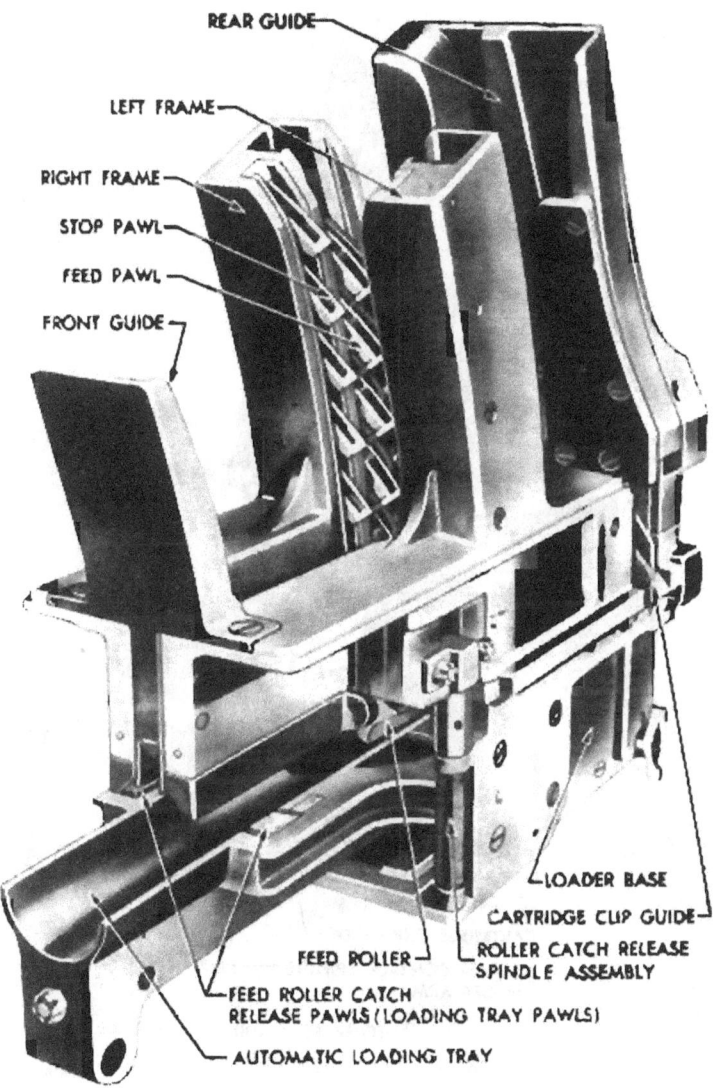

Figure 28 — Automatic Loader — Front View

**TM 9-252**
**14**

**40-MM AUTOMATIC GUN M1 (AA) AND 40-MM ANTIAIRCRAFT GUN CARRIAGES M2 AND M2A1**

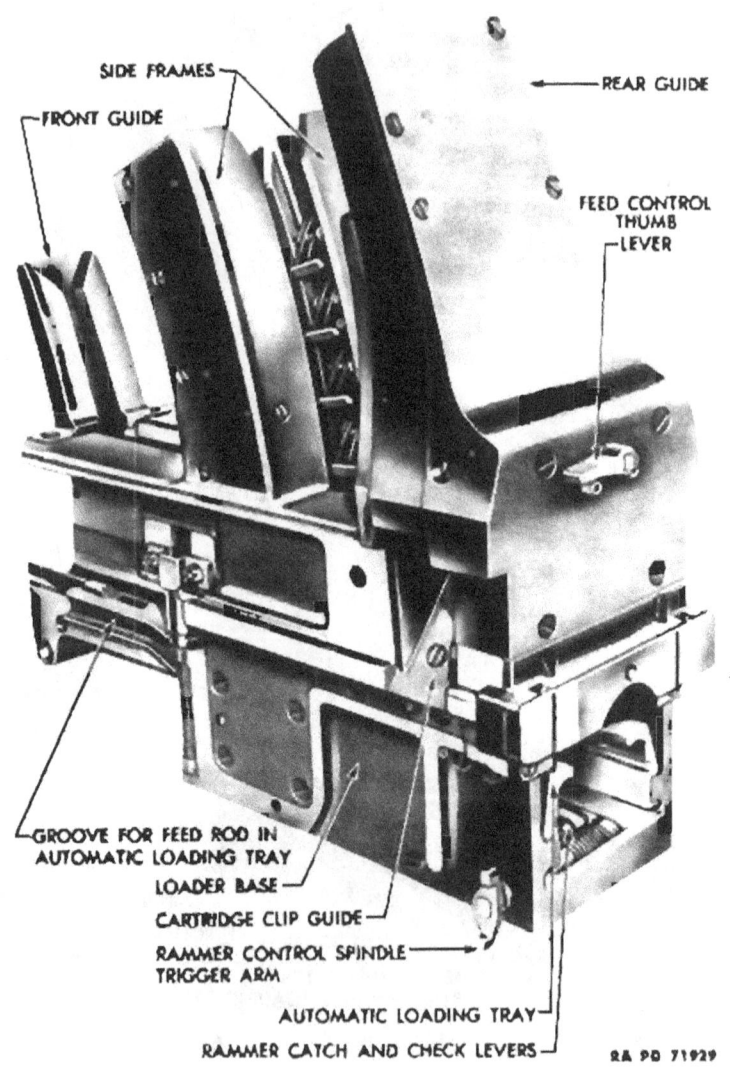

Figure 29 — Automatic Loader — Rear View

TM 9-252
14

## DESCRIPTION AND FUNCTIONING OF GUN

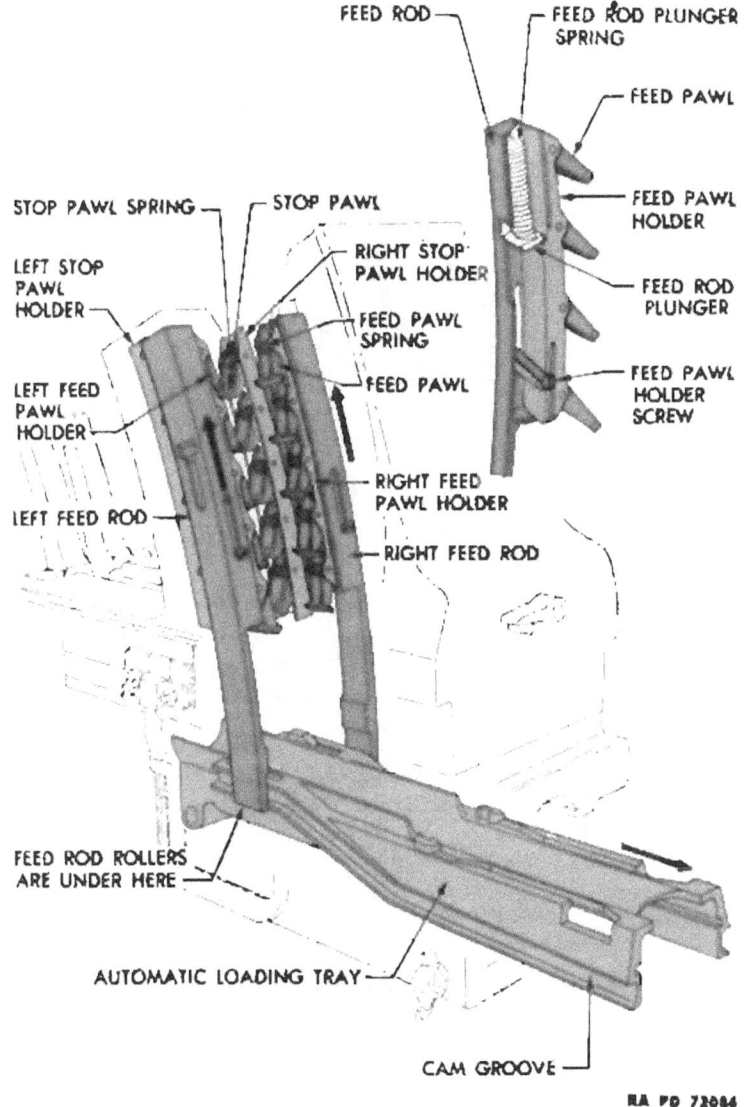

Figure 30 — Feed Mechanism — Raised Position

TM 9-252
14

**40-MM AUTOMATIC GUN M1 (AA) AND 40-MM ANTIAIRCRAFT GUN CARRIAGES M2 AND M2A1**

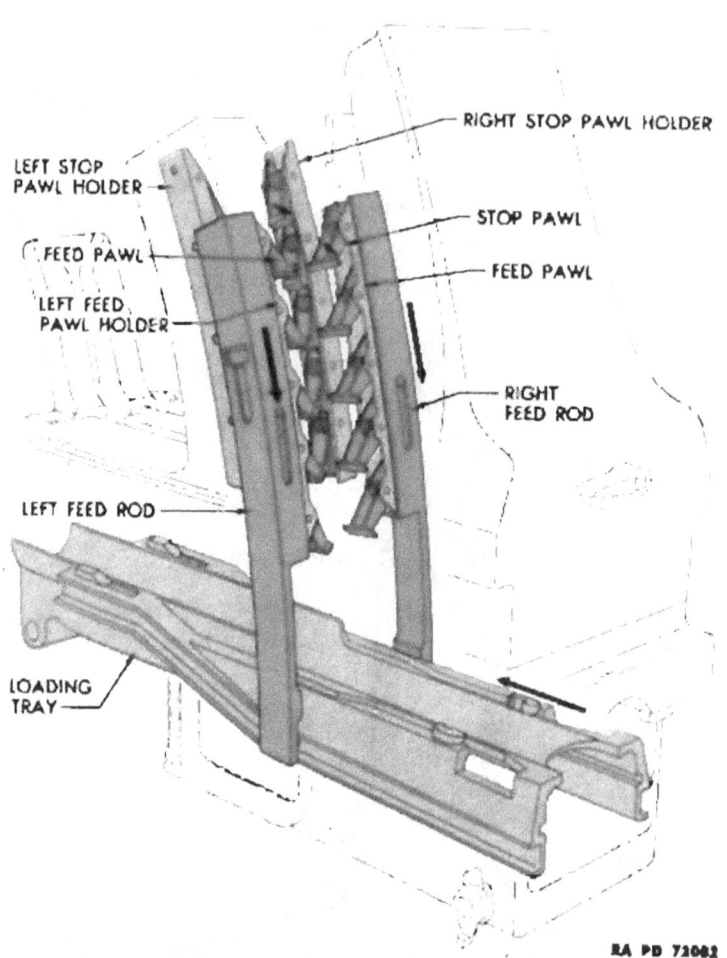

*Figure 31 — Feed Mechanism — Lowered Position*

supply of cartridges. The mechanism consists of the front and rear guides, side frames, feed rods and rollers, feed and stop pawls and holders, feed rollers, catches, and plungers, and the cartridge clip release arrangement.

b. Feed and Stop Pawls.

(1) The feed and stop pawls and holders are housed in the side

## DESCRIPTION AND FUNCTIONING OF GUN

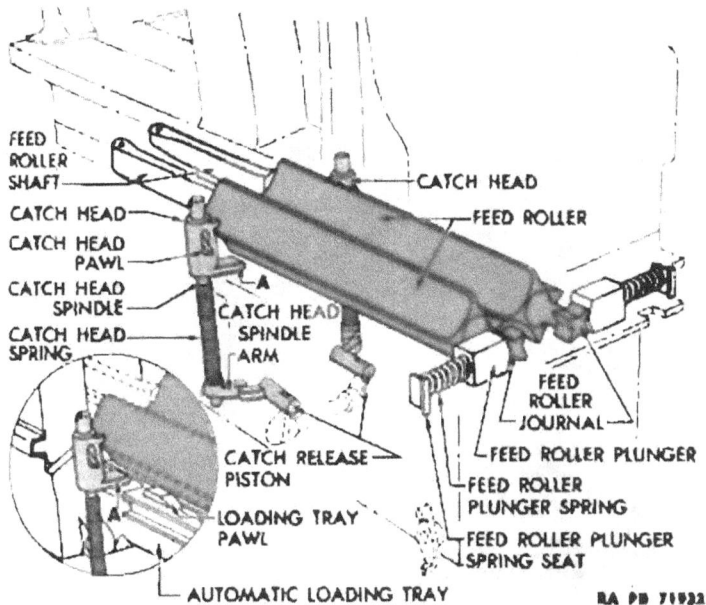

*Figure 32 — Feed Rollers — Engaged*

frames. The feed pawls are operated by the feed rods which are raised and lowered by the feed rod rollers moving in the cam grooves in the sides of the loading tray as the loading tray recoils and counterrecoils. The feed pawls move all cartridges in the loader downward each time a round is fired. The stop pawls, which are retained in stationary holders, prevent the cartridges from moving upward during recoil.

(2) During recoil, the feed rod rollers moving in the cam grooves in the sides of the loading tray raise the feed rods. The feed rods carry the feed pawl holders and pawls upward, the pawls riding against the sides of the cartridges in the loader on the upward movement. The stop pawls are forced outwardly from their holders by their torsion springs and prevent the cartridges from being raised as the feed pawls move upward.

(3) The feed rods draw the feed pawls downward in counterrecoil. The feed pawls, forced outwardly from their holders by their torsion springs, engage the cartridges and force them downward. The stop pawls are forced into their holders by the cartridges, permitting a cartridge to be fed onto the loading tray.

(4) Springs and plungers are provided on the tops of the feed rods

TM 9-252
14

**40-MM AUTOMATIC GUN M1 (AA) AND 40-MM ANTIAIRCRAFT GUN CARRIAGES M2 AND M2A1**

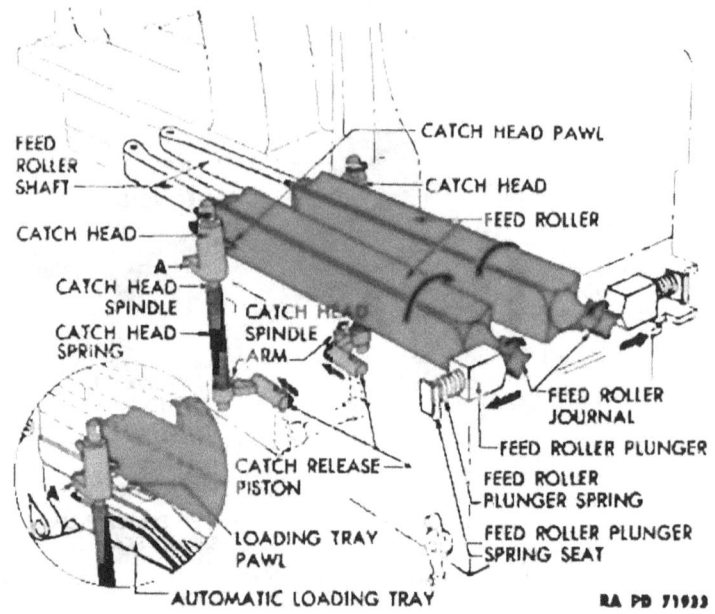

*Figure 33 — Feed Rollers — Released*

to protect the mechanism in the case of a jammed cartridge. Should a cartridge jam during the downward movement, the compression of these springs permits the feed pawl holders to remain in the "up" position.

c. **Feed Rollers.**

(1) The feed rollers (red in figs. 32 and 33) are metal prisms with four convex sides. They are located in the loader under the feed and stop pawl assemblies and over the loading tray. They are mounted to rotate in opposite directions. Their purpose is to insure that only one cartridge is fed onto the loading tray at a time.

(2) The feed rollers are revolved by the cartridge as it is forced down onto the loading tray. Their movement is controlled by the feed roller catch and plunger mechanisms.

d. **Feed Roller Catch and Feed Roller Plunger Mechanisms.**

(1) The feed roller catch mechanisms (green in figs. 32 and 33) consist of feed roller catch heads fitted with spring-loaded catch head pawls, catch head spindles, springs, spindle arms, and catch release pistons. The catch mechanisms are located near the front ends of the feed rollers. Their functions are to hold and release the feed

50

## DESCRIPTION AND FUNCTIONING OF GUN

rollers and to limit their turning to exactly one-quarter revolution. They are actuated by the catch release pawl assemblies in automatic operation (inserts, figs. 32 and 33) and by the catch release pistons in manual operation (figs. 32 and 33).

(2) The feed roller plunger mechanisms consist of feed roller plungers, springs, and spring seats (figs. 32 and 33). They are located under the loader crosspiece cover at the rear of the loader. The plungers act on 4-pointed feed roller journals on the rear ends of the feed rollers to aline the rollers after each one-quarter turn.

(3) The spring-loaded catch release pawl assemblies (figs. 35 and 36) are located at the front of the loading tray. During the recoil of the gun, the loading tray pawls of the catch release pawl assemblies are depressed as they pass under the lugs (A, figs. 32 and 33) which protrude sidewise from the bottoms of the catch heads, but during counterrecoil, they engage these lugs and rotate the catch heads.

(4) The feed rollers are normally locked in position by the catch heads. During counterrecoil, when the cartridges are being depressed by the feed mechanism, the loading tray pawls engage and rotate the catch heads. This action releases the feed rollers and permits them to be revolved by the cartridge as it is forced downward (inserts, figs. 32 and 33). When the cartridge has passed through the feed rollers and the rollers have revolved one-quarter turn, the catch heads are returned to their normal positions by their torsion springs, locking the rollers in place.

(5) The catch heads are also rotated in another manner. When the hand operating lever (fig. 37) is pulled backward, it actuates the hand operating device which rotates the feed roller catch release spindle levers in a forward direction. These move the catch release pistons (figs. 32 and 33) against the catch head spindle arms, rotating the catch heads and releasing the feed rollers for one-quarter turn.

*e.* Cartridge Clip Release Arrangement.

(1) The cartridges are released from the cartridge clip (red in fig. 34) as they move downward in the automatic loader. As the cartridge and clip move downward, the cartridge clip pins are forced to the rear by the cam surface of the lower rail in the rear cartridge guide. This action releases the cartridges from their hooks and from the clip. When fully released, the clip is deflected by the rear rail and is ejected through the cartridge clip guide on the left side of the gun.

(2) The action on the cartridge clip pins is shown in figure 34. Pin A has not engaged the lower rail. Pin B has engaged the lower rail and is just being forced rearward. Pins C and D have been forced fully to the rear and have released their cartridges. The action of the hooks is shown in the insert in figure 34. Hooks E and F hold

TM 9-252
14-15

**40-MM AUTOMATIC GUN M1 (AA) AND 40-MM ANTIAIRCRAFT GUN CARRIAGES M2 AND M2A1**

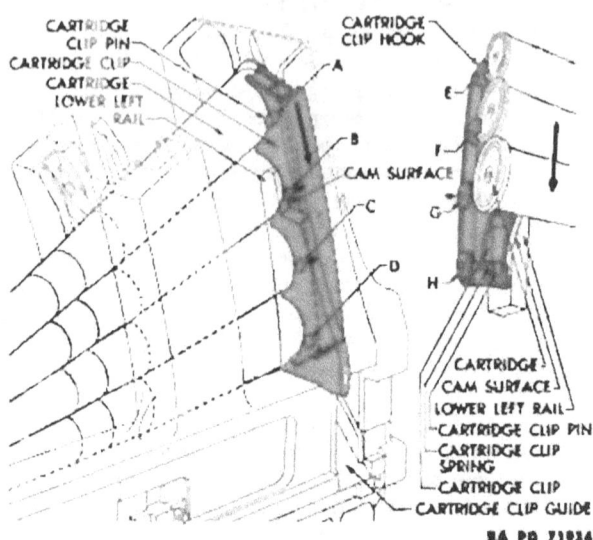

*Figure 34 — Cartridge Clip — Release From Cartridges*

their cartridges. Hook G has been forced to the rear, releasing its cartridge. The cartridge has been removed from hook H.

## 15. AUTOMATIC LOADING TRAY.

a. The functions of the loading tray are to support the cartridge until it is rammed into the chamber by the rammer, to rotate the feed roller catch heads and thus release the feed rollers, to actuate the feed rods, to operate the rammer releasing lever, and to support and actuate the cartridge rammer assembly. These duties are performed by various surfaces and mechanisms either incorporated into the shape of the tray or attached to it.

b. The loading tray (green in figs. 35 and 36) is trough-shaped to receive the cartridge. Its front end is bolted to the breech ring and recoils and counterrecoils with it. The loading tray pawls of the catch release pawl assemblies, which rotate the catch heads, are located near the front end. The cam grooves in which the feed rod rollers move are located on the sides of the tray. These grooves cause the rollers to raise the feed rods during recoil and to force them down during counterrecoil, actuating the feed mechanism and feeding another cartridge onto the loading tray.

c. The housing for the rammer rod and spring is cast in the under side of the tray toward the front. Cam slots are provided along

## DESCRIPTION AND FUNCTIONING OF GUN

both sides in the top of the tray for the rammer levers. A beveled projection on the bottom of the rammer rod and spring housing (A, fig. 35) trips the rammer releasing lever as the loading tray nears the end of counterrecoil.

### 16. CARTRIDGE RAMMER ASSEMBLY.

a. The function of the cartridge rammer assembly (red in figs. 35 and 36) is to store up energy by the compression of its spring during the counterrecoil of the loading tray and, by means of the rammer levers, to grip the rim of the cartridge; then, when the rammer shoe is released, to carry the cartridge forward under the compression of the rammer spring, release it, and send it forward by its momentum into the chamber of the gun.

b. The cartridge rammer rod, head, spring, and seat are contained within the housing in the under side of the automatic loading tray. The rammer shoe is fastened to the rear end of the rammer rod. The rammer levers are pivoted in the sides of the shoe. Spring-loaded plungers force the lever heads inwardly. A leather rammer buffer pad, mounted in the front end of the housing, absorbs the shock as the rammer is driven forward by the rammer spring when the rammer shoe is released.

c. The rammer can be operated manually or automatically. The rammer must be operated manually to prepare for the firing of the initial round or before firing can be resumed if the loader has been permitted to become empty. To avoid permitting the ammunition in the loader to run out, the feed control thumb lever is usually placed in the left position when firing. This will stop automatic loading and firing when only one cartridge remains in the feed guides and one cartridge is on the loading tray.

d. Manual Operation.

CAUTION: Before initial loading, the outer safety lever should be placed in the "SAFE" position.

(1) When the hand operating lever is pulled backward as far as it will go, it causes the rammer cocking levers (par. 17 c) to force the cartridge rammer shoe rearward, compressing the cartridge rammer spring. With the hand operating lever held in the backward position to permit the catch release pistons to rotate the feed roller catch heads, a clip of cartridges is placed in the automatic loader and pushed downward until the feed rollers rotate one-quarter turn and a cartridge moves down onto the automatic loading tray. The rim of the cartridge is engaged in the grooves of the upper arm flanges of the rammer levers.

NOTE: The cartridge rammer shoe is carried farther backward during manual operation than during automatic operation.

(2) When the hand operating lever is replaced in its horizontal

**TM 9-252**
**16**

**40-MM AUTOMATIC GUN M1 (AA) AND 40-MM ANTIAIRCRAFT GUN CARRIAGES M2 AND M2A1**

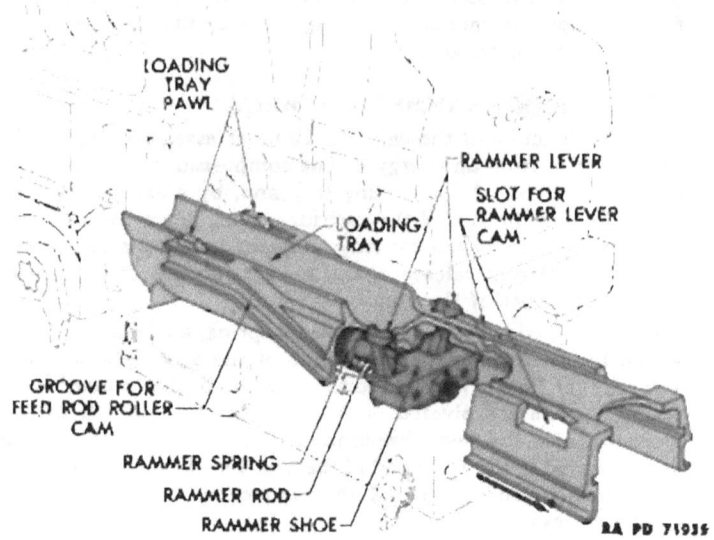

*Figure 35 — Loading Tray — Rammer Shoe Held*

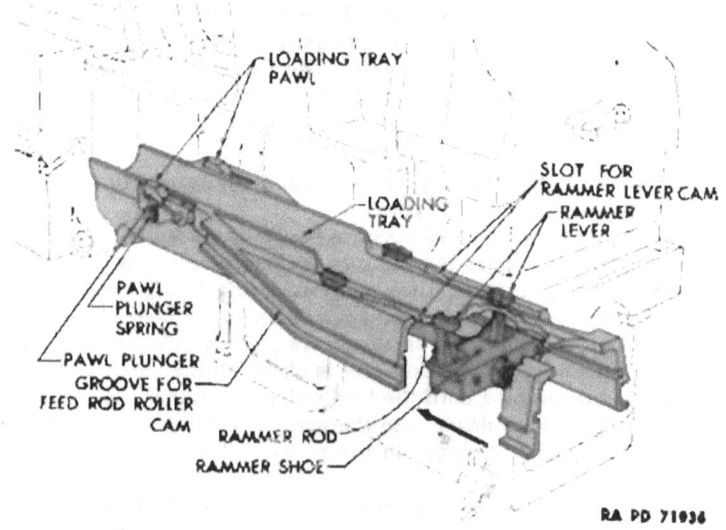

*Figure 36 — Automatic Loading Tray — Rammer Shoe Released*

54

## DESCRIPTION AND FUNCTIONING OF GUN

position, the cocking levers will return to their normal positions and the rammer shoe will move forward a short distance. Here the rammer shoe will be engaged by the left check lever (par. 17 f), the rammer catch lever being depressed by the rammer releasing lever as the result of the projection on the under side of the loading tray depressing the toe of the rammer releasing lever.

(3) If the check levers are depressed, the rammer shoe is carried forward by the action of the compressed rammer spring. When the rammer shoe moves forward, the rammer levers carry the cartridge with it. Near the end of the forward motion of the rammer shoe, the rammer levers are forced outwardly by the cam slots in the loading tray, releasing the cartridge which is thrown into the chamber of the gun by the momentum of the ramming action.

e. **Automatic Operation.**

(1) During recoil, the rammer shoe assembly is carried rearward with the loading tray. As the loading tray starts forward in counterrecoil, the lower portion of the rammer shoe is engaged by the catch lever. This lever holds the rammer shoe in its rearward position while the loading tray continues to return to battery with the gun and breech ring, causing the rammer spring to become compressed.

(2) As the loading tray nears the end of counterrecoil, the beveled projection on the under side of the loading tray trips the rammer releasing lever, freeing the rammer shoe from the restraint of the catch lever.

(3) If the catch check levers are depressed, the rammer shoe is carried forward by the action of the compressed rammer spring to ram the cartridge and fire the gun.

## 17. AUTOMATIC LOADER CONTROL MECHANISM.

a. The automatic loader control mechanism consists of those parts of the automatic loader which cock, hold, and release the cartridge rammer shoe and which actuate the catch heads which hold and release the feed rollers. The feed roller control parts are described in paragraph 14 d.

b. The rammer control parts consist of: the rammer cocking lever shaft assembly (also known as the hand operating device) which is actuated by the hand operating lever to compress the rammer spring manually; the catch and release mechanism which holds the rammer shoe while the spring is being compressed automatically and insures that the rammer shoe will not be released until the gun has reached the end of counterrecoil; the feed control check and release mechanism which can be set to stop automatic loading and firing when only one cartridge remains in the feed guides; and the firing mechanism check and release mechanism which is controlled by the outer safety lever.

**TM 9-252**
**17**

## 40-MM AUTOMATIC GUN M1 (AA) AND 40-MM ANTIAIRCRAFT GUN CARRIAGES M2 AND M2A1

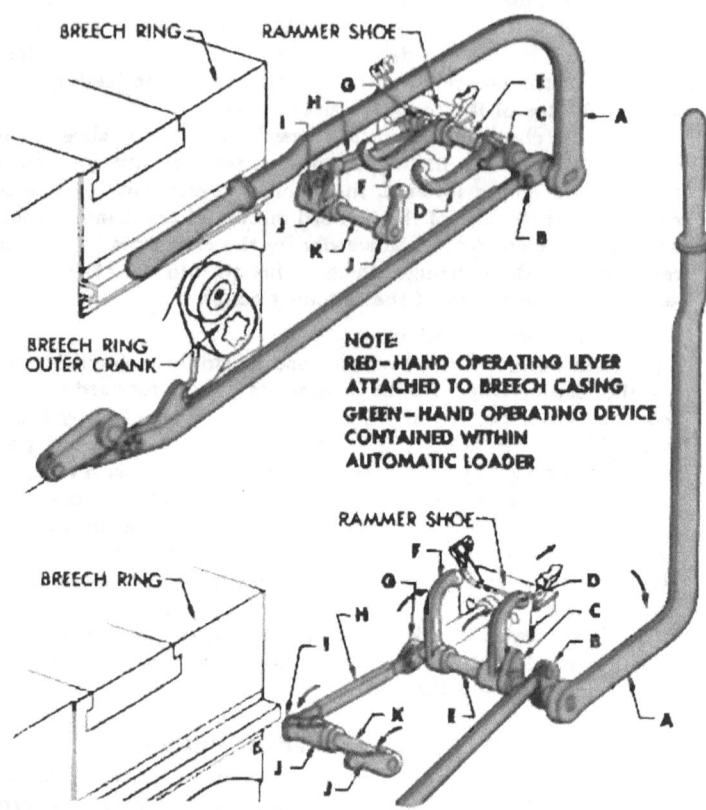

A—HAND OPERATING LEVER
B—HAND OPERATING LEVER SHAFT
C—HAND OPERATING LEVER SHAFT ARM
D—LEFT RAMMER COCKING LEVER
E—RAMMER COCKING LEVER SHAFT
F—RIGHT RAMMER COCKING LEVER
G—RAMMER COCKING LEVER SHAFT ARM
H—FEED ROLLER CATCH RELEASE LINK
I—FEED ROLLER CATCH RELEASE SPINDLE ARM
J—FEED ROLLER CATCH RELEASE SPINDLE LEVER
K—FEED ROLLER CATCH RELEASE SPINDLE

*Figure 37 — Hand Operating Lever and Hand Operating Device*

TM 9-252
17

## DESCRIPTION AND FUNCTIONING OF GUN

**c. Rammer Cocking Lever Shaft Assembly (Hand Operating Device).**

(1) This mechanism (fig. 37) is housed in the base of the automatic loader. It is operated manually by the hand operating lever to force the rammer shoe rearward and compress the rammer spring. At the same time, it presses the feed roller catch release pistons forward to contact the catch head spindle arms and rotate the feed roller catch heads. This frees the feed rollers for a one-quarter turn resulting from pressure applied on the cartridges in the magazine.

(2) The functioning of the mechanism is shown in detail in figure 37 (green). When the hand operating lever A is pulled backward, the hand operating lever shaft arm C engages the left rammer cocking lever D. The rammer cocking lever shaft E and the left and right rammer cocking levers D and F are rotated backward, carrying the cartridge rammer shoe rearward.

(3) The rammer cocking lever shaft E also rotates the rammer cocking lever shaft arm G, transmitting motion through the feed roller catch release link H and catch release spindle arm I to the catch release spindle levers J. These move the catch release pistons forward to contact the catch head spindle arms and rotate the feed roller catch heads.

**d. Automatic Catch and Release Mechanism.**

(1) The rammer catch lever of the automatic catch and release mechanism (uncolored in fig. 38) is the central lever of the three mounted on the rammer catch lever axis pin in the rear of the automatic loader base. Its spring-loaded plunger forces its rear end upward to engage the shoulder on the under side of the rammer shoe.

(2) If the loader is adequately charged with ammunition, the outer safety lever is set for automatic fire, and one of the firing pedals is held in the depressed position, the rammer check lever is the only one which holds the rammer shoe until the rammer spring is compressed. This lever is not released until the gun is near the end of counterrecoil.

(3) This lever is released by the beveled projection on the bottom of the loading tray engaging the front end of the rammer releasing lever when the loading tray is near the end of counterrecoil. Then, the rammer releasing lever pivots on its axis pin and its rear end lifts the front end of the rammer catch lever. The rear end of the rammer catch lever is forced downward against its spring-loaded plunger, releasing the rammer shoe. The rammer shoe is driven forward by the rammer spring unless the shoe is held by one or both check levers.

**e. Feed Control Check and Release Mechanism.**

(1) The function of the feed control check and release mechanism is to stop automatic loading and firing when only one cartridge remains in the feed guides and one cartridge is on the loading tray. This

TM 9-252
17

**40-MM AUTOMATIC GUN M1 (AA) AND 40-MM ANTIAIRCRAFT GUN CARRIAGES M2 AND M2A1**

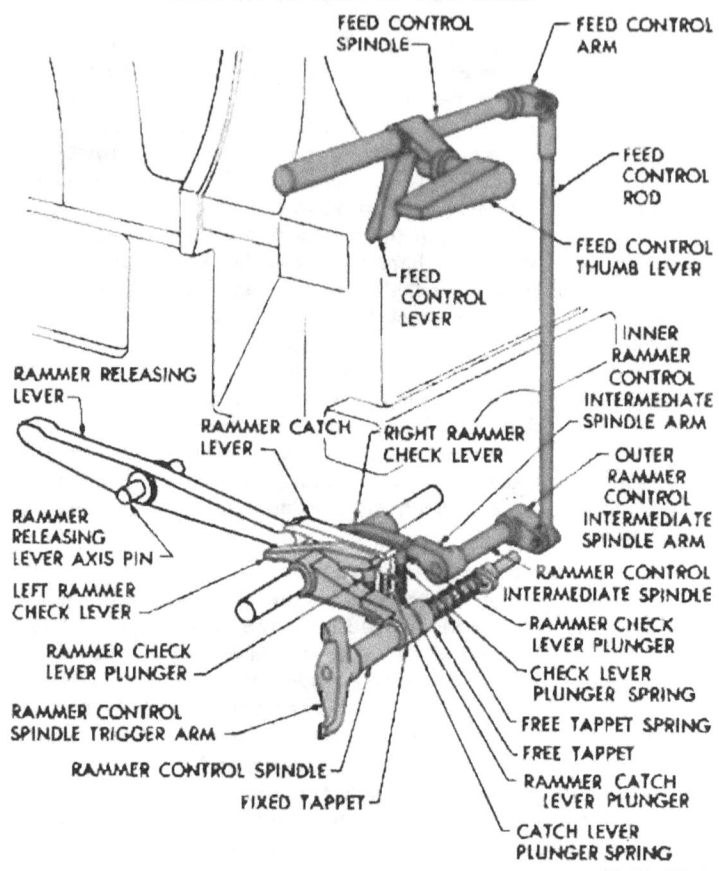

*Figure 38 — Catch and Release Mechanism — All Levers Depressed*

eliminates the necessity of reloading by hand as must be done if the supply of ammunition in the loader runs out. The mechanism operates when the feed control thumb lever is pointed to the left. The mechanism may be made inoperative to eliminate the reserve feature by pointing the feed control lever to the right.

(2) The check lever of the feed control check and release mechanism (red in fig. 38) is the right lever of three mounted on the rammer catch lever axis pin. Its rear end is raised by a spring-loaded plunger to engage the shoulder of the rammer shoe unless the lever is held in the depressed position. The lever has an integral arm which is contacted by the inner rammer control intermediate spindle arm.

58

## DESCRIPTION AND FUNCTIONING OF GUN

(3) The feed control check and release mechanism is actuated by the feed control lever at the rear and inside the automatic loader and by the feed control thumb lever at the rear and on the outside of the rear guide of the loader. Both levers act upon the inner rammer control intermediate spindle arm through the feed control spindle, arm, rod, outer rammer control intermediate spindle arm, and rammer control intermediate spindle to depress the rear end of the right check lever.

(4) The feed control check and release mechanism is controlled also by the feed control thumb lever. When this lever is in the left-hand position, the mechanism is actuated by the cartridges. It is placed in this position for single and automatic fire. When more than one cartridge is in the feed guides of the loader, the feed control lever is held to the rear of the loader by the base of the second cartridge above the feed rollers. In this case, the inner rammer control intermediate spindle arm holds the right check lever down where it has no effect on the holding of the rammer shoe.

(5) When only one round is left above the feed rollers, the feed control lever is free to lift and no longer has a restraining influence on the inner rammer control intermediate spindle arm. The rear end of the right check lever is raised by its spring-loaded plunger and holds the rammer shoe in its rearward position until the check lever is depressed. As soon as more cartridges are inserted in the loader, the releasing mechanism comes into operation again and automatic firing may be resumed.

(6) When the lever is in the right-hand position, the eccentric on the shaft of the feed control thumb lever engages and operates the short arm of the feed control lever, and through the linkage, causes the rear end of the right check lever to be depressed and the reserve feature is eliminated. This action is also utilized in releasing the rammer shoe when the weapon is unloaded.

f. Firing Mechanism Check and Release Mechanism.

(1) This mechanism (green in fig. 38) is controlled by the outer safety lever. In single fire operation, it is intended to be the last link in the holding of the rammer shoe prior to its release and the firing of the gun. It is released when the outer safety lever is set either for single fire or automatic fire and the firing pedal is depressed.

(2) The check lever of the firing mechanism check and release mechanism is the left lever of three mounted on the rammer catch lever axis pin. Its rear end is raised by a spring-loaded plunger to engage the rammer shoe unless the lever is held in the depressed position. This check lever has an integral arm which is contacted by the fixed tappet keyed to the rammer control spindle.

(3) When the outer safety lever is properly set and the firing pedal is depressed, the firing lever pawl of the breech casing firing mechanism engages and rotates the trigger and the rammer control

TM 9-252
17-18

**40-MM AUTOMATIC GUN M1 (AA) AND 40-MM ANTIAIRCRAFT GUN CARRIAGES M2 AND M2A1**

spindle trigger arm. This arm rotates the rammer control spindle and fixed tappet. The tappet depresses the firing mechanism check lever, releasing the rammer shoe, if this lever is the only one engaging the rammer shoe.

(4) When the firing lever pawl disengages the trigger, the left check lever is raised to its normal position by the action of its spring-loaded plunger. The free tappet is held against the fixed tappet by the free tappet spring, and snubs its action.

## 18. FIRING CYCLE, AUTOMATIC FIRE.

a. The motions of the various parts of the gun occur in a definite and interrelated manner during the firing cycle. To bring out the relationship of the parts and to illustrate their functions at specific points during the firing cycle, six drawings have been prepared (figs. 39 through 44) which show the actions and positions of the affected parts at six stages during the cycle.

b. To obtain automatic fire, certain conditions must be met. More than two cartridges must be above the feed rollers to keep the feed control check lever depressed. The feed control thumb lever must be in the left position so it will be unnecessary to reload by hand. The outer safety lever must be set on "AUTO FIRE" and one of the firing pedals must be held in the engaged position. With these conditions met, the right and left rammer check levers remain depressed and the rammer catch lever is the only one that can restrain the rammer shoe.

c. Part of the firing cycle occurs during recoil; the balance occurs during and at the end of counterrecoil. In the illustrations (figs. 39 through 44), the recoiling parts of the weapon are shown in red; the automatic loader is shown in green. Figure 45 shows the time required for the various parts of the gun to perform their functions during recoil and counterrecoil.

d. Recoil.

(1) FIRST STAGE. Figure 39 shows the position of the parts just after the percussion cap of the cartridge has been struck by the firing pin. Immediately after firing, the tube, breech ring, and loading tray begin to recoil and the projection on the bottom of the loading tray relieves its pressure from the rammer releasing lever, permitting the rammer catch lever to raise. During the first few inches of recoil, the outer crank rotates enough to cause the firing pin to be withdrawn into the breechblock.

(2) As recoil continues, the outer crank continues to be rotated by the cam surfaces of the side cover. The inner cranks begin to lower the breechblock. In doing so, they actuate the outer cocking lever which cocks the firing pin for the next shot. The check plunger engages the inner cocking lever.

TM 9-252
18

## DESCRIPTION AND FUNCTIONING OF GUN

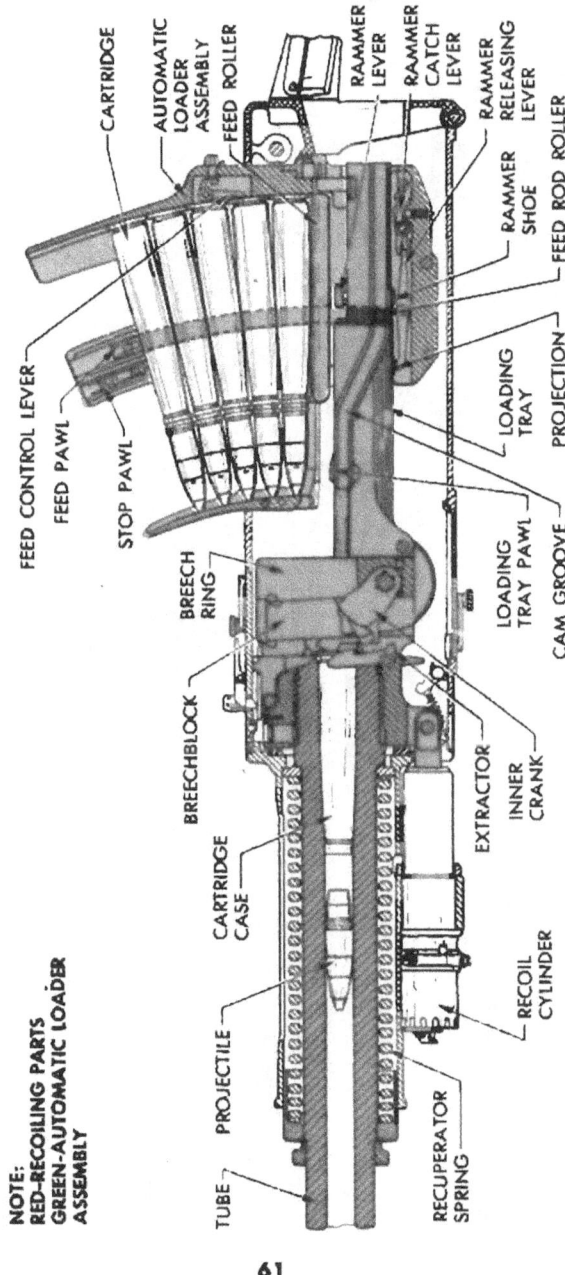

*Figure 39 — Automatic Firing Cycle — First Stage — Gun Firing*

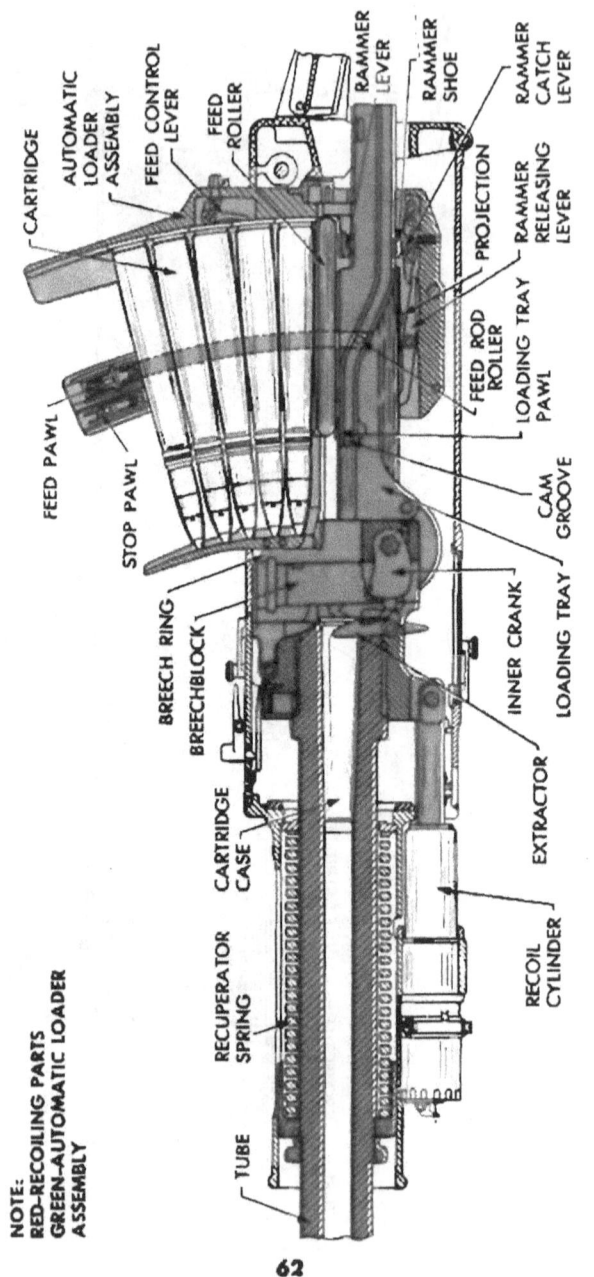

Figure 40 — Automatic Firing Cycle — Second Stage — Breechblock Being Lowered

**TM 9-252**
**18**

## DESCRIPTION AND FUNCTIONING OF GUN

(3) As the loading tray moves backward, the cam grooves move past the rollers on the ends of the feed rods and the rollers enter the inclined portion of the grooves, raising the feed rods. The feed rods carry the feed pawls upward and over the next cartridge in the loader. The cartridges are prevented from being raised by the stop pawls. The loading tray pawls on the front of the loading tray are depressed by and pass under the lugs which extend sidewise from the feed roller catch heads.

(4) SECOND STAGE. Figure 40 shows the position of the parts when the breechblock has been lowered part of the way in its slides in the breech ring. The feed rods, holders, and pawls are nearing their extreme upward position. The rammer shoe is over the rammer catch lever. The loading tray is free of obstructions to permit ejecting the empty case because the rammer levers were forced to the sides of the tray by their cam grooves at the same time they released the cartridge in ramming it.

(5) As the breechblock descends, the projections at the sides of the front face of the breechblock strike the toes of the extractors. The extractors are rotated toward the rear and their lips catch the rim of the cartridge case and eject it. The empty case is thrown with considerable force backward along the loading tray, through the rear cover opening, against the cartridge case deflector, and into the cartridge chute. The case is carried down the chute, under the gun, and out in front of the weapon.

(6) THIRD STAGE. Figure 41 shows the position of the parts at the end of recoil. The empty cartridge case has been ejected. The breechblock is in its lowered position. The feed rod rollers are in the upper horizontal portion of the loading tray cam grooves and the feed rods are fully raised. Recoil has been stopped by the action of the recuperator spring and the recoil cylinder. The hook-shaped heads of the extractors are over the notched tops of the projections on the front face of the breechblock. The rammer shoe is to the rear of the rammer catch lever, and the loading tray pawls are to the rear of the feed roller catch heads.

e. **Counterrecoil.**

(1) FOURTH STAGE. The tube, breech ring, and loading tray start to move into battery. The breechblock moves slightly upward under the action of the closing spring until it is brought to a stop and held by the extractor heads engaging the notches in the tops of the breechblock projections. The rammer shoe is engaged by the rammer catch lever, holding the shoe and compressing the rammer spring as the tube, breech ring, and loading tray move into battery (fig. 42). The right check lever is held out of engagement because there are sufficient cartridges in the loader to hold the feed control lever in its rearward position. The left check lever is held out of engagement be-

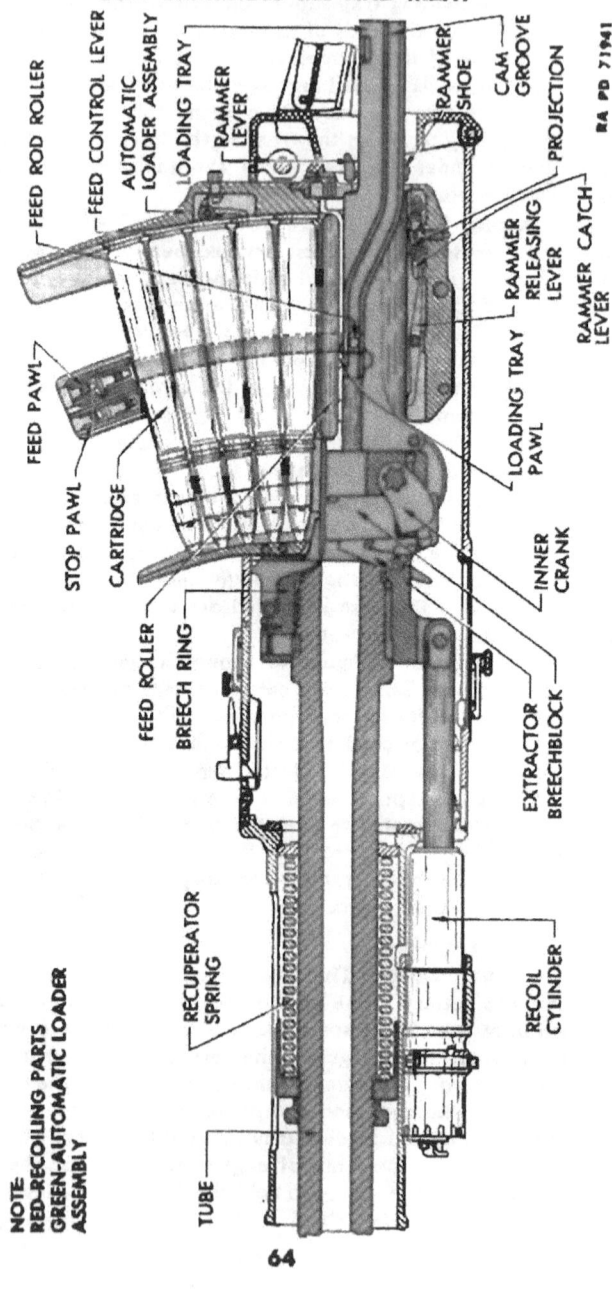

Figure 41 — Automatic Firing Cycle — Third Stage — Breechblock Lowered, Cartridge Ejected

TM 9-252
18

## DESCRIPTION AND FUNCTIONING OF GUN

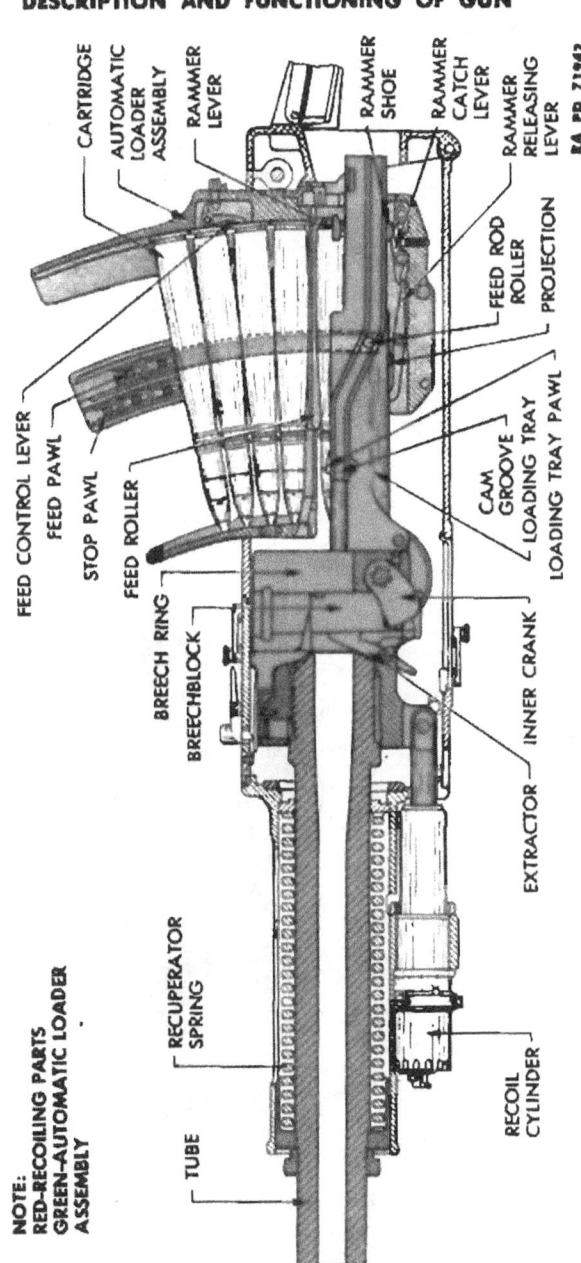

*Figure 42 — Automatic Firing Cycle — Fourth Stage — Cartridge on Loading Tray*

# TM 9-252
## 40-MM AUTOMATIC GUN M1 (AA) AND 40-MM ANTIAIRCRAFT GUN CARRIAGES M2 AND M2A1

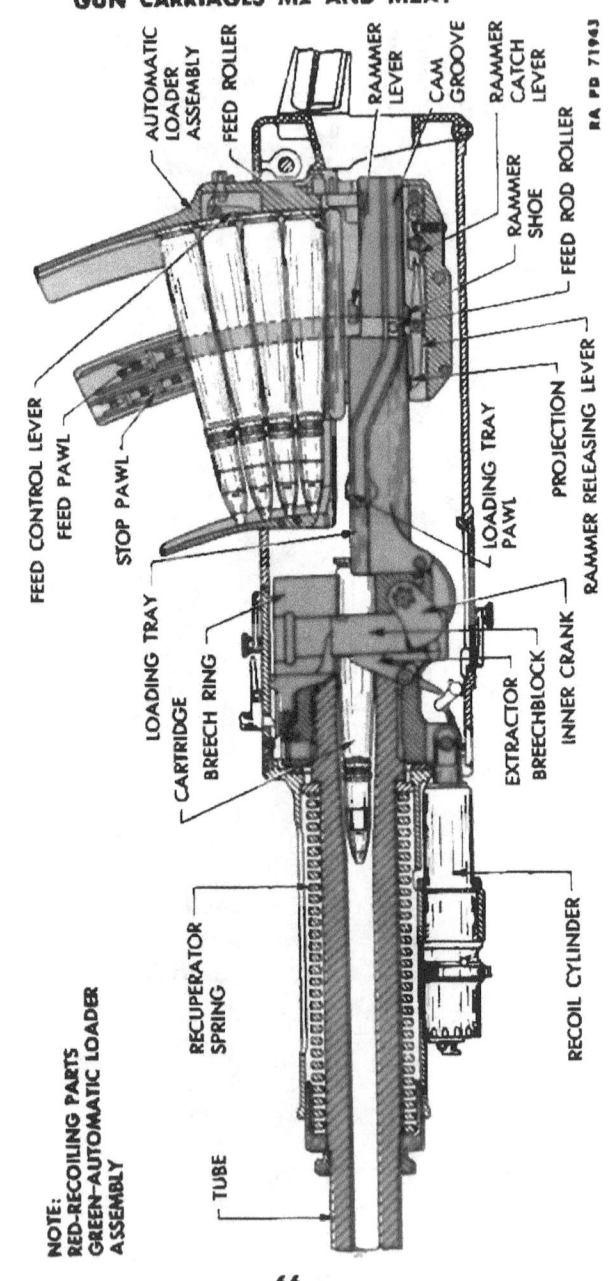

Figure 43 — Automatic Firing Cycle — Fifth Stage — Cartridge Being Rammed

## DESCRIPTION AND FUNCTIONING OF GUN

cause one of the firing pedals is depressed and the outer safety lever is set for automatic fire.

(2) As the loading tray moves forward, the loading tray pawls on the front end of the tray engage the lugs on the feed roller catch heads, rotating the catch heads, and releasing the feed rollers for one-quarter turn. At the same time, the feed rod rollers enter the declined portion of the cam grooves on the loading tray, forcing the feed rods and pawls downward. The feed pawls engage the cartridges, forcing them downward.

(3) The lowest cartridge rotates the feed rollers, passes through them, and drops on the loading tray into the rammer shoe levers. The pawls on the feed roller catch heads engage the feed rollers, preventing them from revolving more than one-quarter turn, thus preventing more than one cartridge from passing through. The cam slots in top of the loading tray cause the heads of the rammer levers to be forced inwardly to grip the rim of the cartridge. Figure 42 shows the position of the parts as the cartridge drops onto the loading tray and its rim is engaged by the rammer levers.

(4) FIFTH STAGE. When the feed rollers have completed a quarter turn, they are relocked by the catch heads which are returned to their normal positions by their torsion springs. As the gun nears the end of counterrecoil, the beveled projection on the bottom of the loading tray trips the rammer releasing lever, freeing the rammer shoe from the restraint of the rammer catch lever.

(5) The rammer shoe is pulled forward by the rammer spring, the rammer levers carrying the cartridge forward with the shoe. As the rammer shoe nears the end of its travel, the cam slots in the tops of the loading tray force the rammer levers outward, releasing the cartridge. The cartridge is thrown forward through the U-shaped channel in the top of the breechblock and into the chamber of the gun. Figure 43 shows the position of the parts as the cartridge enters the chamber.

(6) SIXTH STAGE. After the rim of the cartridge passes through the U-shaped channel at the top of the breechblock, it engages the extractors, pulling them forward, and releasing the breechblock. The closing spring forces the breechblock upward. The beveled front surface of the breechblock engages the rear of the cartridge, forcing it completely into the chamber.

(7) As the breechblock reaches its uppermost position, the projection on the right inner crank contacts the beveled end of the check plunger, moving the plunger to the left, releasing the inner cocking lever, and permitting the firing pin to be thrust forward by the firing pin spring. The cartridge is fired, starting the cycle over again. Figure 44 shows the position of the parts at the instant before the cartridge is fired.

TM 9-252
18

**40-MM AUTOMATIC GUN M1 (AA) AND 40-MM ANTIAIRCRAFT GUN CARRIAGES M2 AND M2A1**

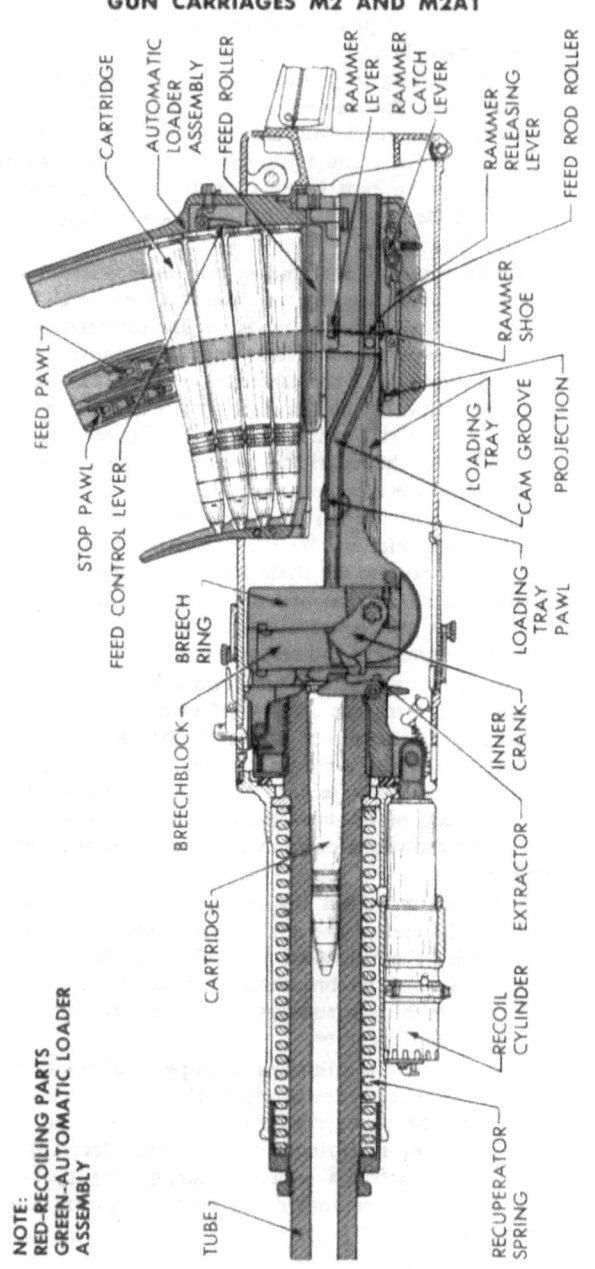

Figure 44 — Automatic Firing Cycle — Sixth Stage — Cycle Completed, Gun in Act of Firing

## DESCRIPTION AND FUNCTIONING OF GUN

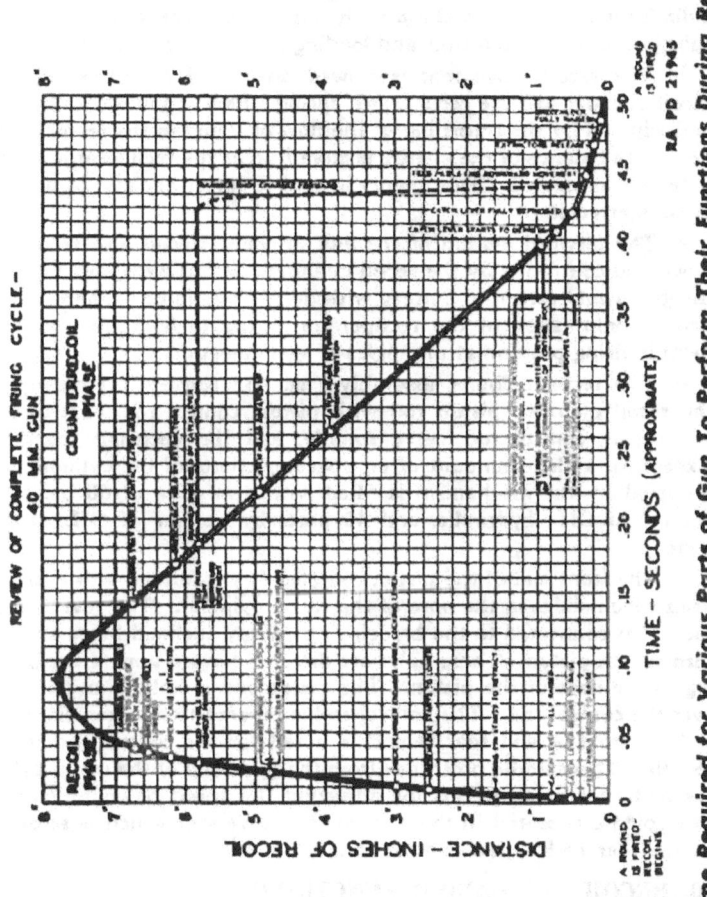

Figure 45 — Time Required for Various Parts of Gun To Perform Their Functions During Recoil and Counterrecoil

## 40-MM AUTOMATIC GUN M1 (AA) AND 40-MM ANTIAIRCRAFT GUN CARRIAGES M2 AND M2A1

**19. RECOIL MECHANISM.**

a. The recoil mechanism consists of the recuperator spring (fig. 6) which is held in compression around the breech end of the tube, and the recoil cylinder assembly (fig. 5) which is mounted under the tubular front end of the breech casing. The recoil rod of the recoil cylinder is connected to the breech ring. When the gun is fired, the tube, recoil rod, breech ring, and loading tray move rearward in recoil.

b. The recoil movement is slowed down and controlled by the recoil cylinder and the recuperator spring. Part of the force of recoil is dissipated by the throttling of the flow of liquid in the recoil cylinder. The balance of the energy is stored up as the recuperator spring is being compressed. This energy is used to return the gun to battery in counterrecoil.

c. The forward motion of the gun in counterrecoil also is slowed down and controlled by the recoil cylinder. If this were not the case, the gun would return to battery with destructive force. There is sufficient compression of the recuperator spring to hold the recoiling parts in firing position at all angles of gun elevation.

d. The recoil cylinder assembly (fig. 46) consists principally of the recoil cylinder, piston rod with piston, control rod, control rod valve seat, control rod valve spindle, and the necessary packing. Except for a slight amount of air space, all parts of the cylinder not occupied by the mechanism is filled with a mixture of 60 parts by volume of OIL, hydraulic, and 40 parts by volume of OIL, recoil, light.

e. The front end of the piston rod (fig. 46) is formed into a piston head which slides in the bore of the recoil cylinder. The rear end of the rod is connected to the breech ring. Eight inclined holes are cut through the piston. These holes are directed toward a metering bushing in the nose of the piston. The piston rod is hollow and travels over the control rod. The control rod is tapered to a larger diameter at both front A and rear B and is bored for the control rod valve spindle. This spindle protrudes from the front end of the recoil cylinder and provides a means for adjusting the speed of counterrecoil. The spindle is seated in the control rod valve seat which is screwed on the rear end of the control rod.

**20. RECOIL MECHANISM FUNCTIONING.**

a. Recoil.

(1) During recoil, the piston rod is drawn to the rear by the recoiling parts of the gun (upper view, fig. 46). Oil in the rear of the piston head is forced through the eight inclined holes C in the piston head, through the metering bushing B and into the space in front of the metering bushing.

## DESCRIPTION AND FUNCTIONING OF GUN

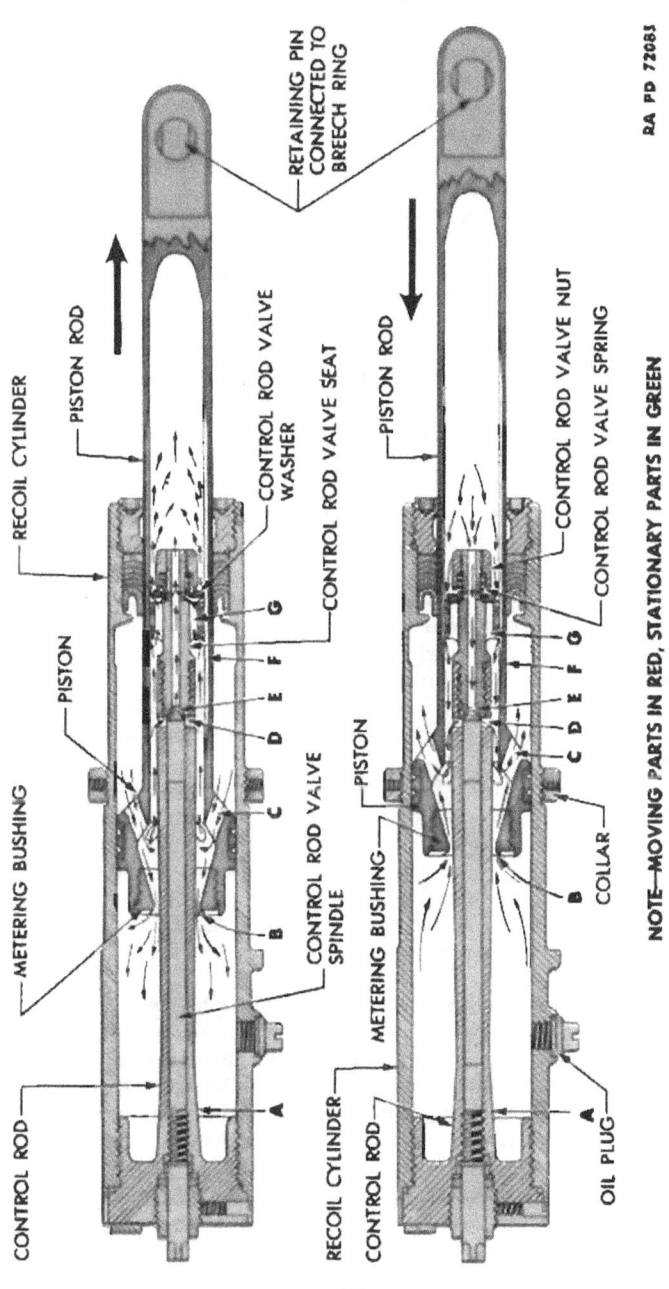

*Figure 46 — Recoil Cylinder — Action in Recoil and Counterrecoil*

## 40-MM AUTOMATIC GUN M1 (AA) AND 40-MM ANTIAIRCRAFT GUN CARRIAGES M2 AND M2A1

(2) The control rod, over which the metering bushing travels, is tapered to a larger diameter at both front and rear. The effective flow space between the control rod and the metering bushing gradually diminishes because of the increasing diameter of the control rod. This offers a greater resistance to the oil as it moves from the rear to the front of the piston.

(3) During recoil, a quantity of oil flows backward between the control rod and the piston rod into the hollow rear end of the piston rod. This oil is used to control counterrecoil. Part of this oil flows through four holes G in the control rod valve seat, forces the control rod valve washer against its spring, and continues on into the rear end of the hollow piston rod. More oil flows around the control rod valve seat through the two tapered grooves F in the walls of the piston rod. A small amount of oil also flows through four radial holes D in the control rod, past the end of the control rod valve spindle, and through the bore E of the control rod valve seat.

(4) The action of the oil in the recoil cylinder, as well as the compression of the recuperator spring, absorbs the energy of the recoil, and slows down and controls the rearward movement of the gun.

h. Counterrecoil.

(1) When recoil ceases, the recuperator spring reasserts itself and forces the tube, breech ring, and loading tray forward, carrying with them the recoil cylinder piston rod and piston. During this forward motion, the oil in front of the piston returns to the rear, flowing between the control rod and the metering bushing (B in lower view, fig. 46) and back through the eight inclined holes in the piston C.

(2) This flow of oil is too rapid to ease the gun back into battery without shock. Full control of counterrecoil is brought about by the restricted release of the oil in the hollow rear end of the piston rod.

(3) As this oil attempts to escape to the front of the recoil cylinder at the beginning of counterrecoil, the control rod valve spring forces the control rod valve washer forward, closing the holes G in the valve seat and preventing the oil back of the valve seat from returning by this path.

(4) The greater part of the oil escapes by flowing through the tapered grooves F in the inside walls of the piston rod. A controlled amount of oil, however, escapes past the pointed end of the control rod valve spindle. Adjustment of this spindle controls the amount of oil which can flow through the bore E in the valve seat to the four radial holes D in the control rod. This adjustment regulates the speed of counterrecoil.

(5) As the gun nears the end of counterrecoil, the two tapered grooves F gradually reduce the flow space around the valve seat. Finally the flow of oil is restricted to that passing the point of the

## DESCRIPTION AND FUNCTIONING OF CARRIAGE

control rod valve spindle. The control rod valve spindle must be precisely adjusted for proper counterrecoil action. The reverse taper A at the front end of the control rod reduces the flow space between the control rod and the metering bushing at this point. Final buffing is accomplished by dash pot action as the piston enters its seat in the rear portion of the head of the control rod.

## Section III
## DESCRIPTION AND FUNCTIONING OF CARRIAGE

### 21. GENERAL.

a. The 40-mm Gun Carriages M2 and M2A1 (AA) are of the 2-axle, 4-wheel trailer type. A drawbar with a standard drawbar lunette forms the connection between the carriage and the prime mover. The carriage is equipped with 4-wheel electric brakes operated from the prime mover, and manually operated mechanical brakes on the rear wheels. The lighting equipment includes taillight, stop, and blackout lights.

b. The top carriage supports the gun, loading platform, firing mechanism, and all parts of the elevating and traversing mechanism which travel in azimuth with the gun. The top carriage can be traversed and the gun can be elevated either manually or electrically. Traverse is continuous. The gun may be depressed and elevated from minus 6 degrees to plus 90 degrees (the elevating limit switch limits power elevation to approximately plus 85 degrees and depression to approximately zero degree).

c. Independent spring suspension is used on all wheels. Axles are spring-loaded to assist in lowering the carriage to firing position and in raising it to traveling position. In firing position, the carriage is leveled by means of four built-in leveling jacks and is anchored by four stakes driven into the ground.

### 22. TOP CARRIAGE.

a. The top carriage (fig. 47) consists of two built-up top carriage frames welded to a rigid cast top carriage frame base. The frames are connected near the center by a large cross tube which also houses the elevating mechanism worm wheel pinion. At the tops of the frames and back of center are the top carriage trunnion bearings and caps which support the tipping parts of the weapon. At the angular points at the front of the frames are the trunnion bearings for the equilibrators.

b. The right top carriage frame carries the traversing gear mecha-

TM 9-252
22-24

**40-MM AUTOMATIC GUN M1 (AA) AND 40-MM ANTIAIRCRAFT GUN CARRIAGES M2 AND M2A1**

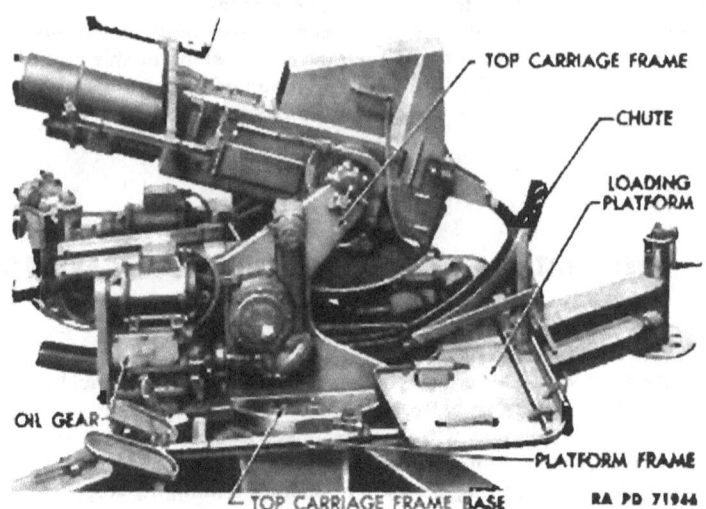

*Figure 47 — Top Carriage and Loading Platform Frame*

nism and the elevating limit switch. The elevating gear mechanism is mounted on and through the left top carriage frame.

c. The loading platform frame is carried on the top carriage frame base. Two levels, used to level the carriage for action, are also mounted on the left side of top carriage frame base.

d. The top carriage rotates in azimuth on the top carriage ball or roller bearing and is driven by the traversing gear assembly.

## 23. PLATFORM FRAME ASSEMBLY.

a. The loading platform frame (fig. 47) is a tubular frame of rectangular shape with two intermediate cross members. These cross members fit the cradles into lugs and are also attached to the lugs on the top carriage frame base.

b. The side members carry the seats and footrests for the elevating and traversing mechanism operators. These seats are adjustable, fore and aft, up and down. The loading platform covers the rear portion of the frame. The oil gears are mounted on plates carried on the two front cross members. The cartridge chute, azimuth indicator, and power synchronizing mechanism (slewing handle) are supported by the platform and frame.

## 24. ELEVATING MECHANISM.

a. The elevating mechanism (fig. 48) consists of the hand elevating mechanism group assembly and the elevating worm and worm

74

TM 9-252
24

**DESCRIPTION AND FUNCTIONING OF CARRIAGE**

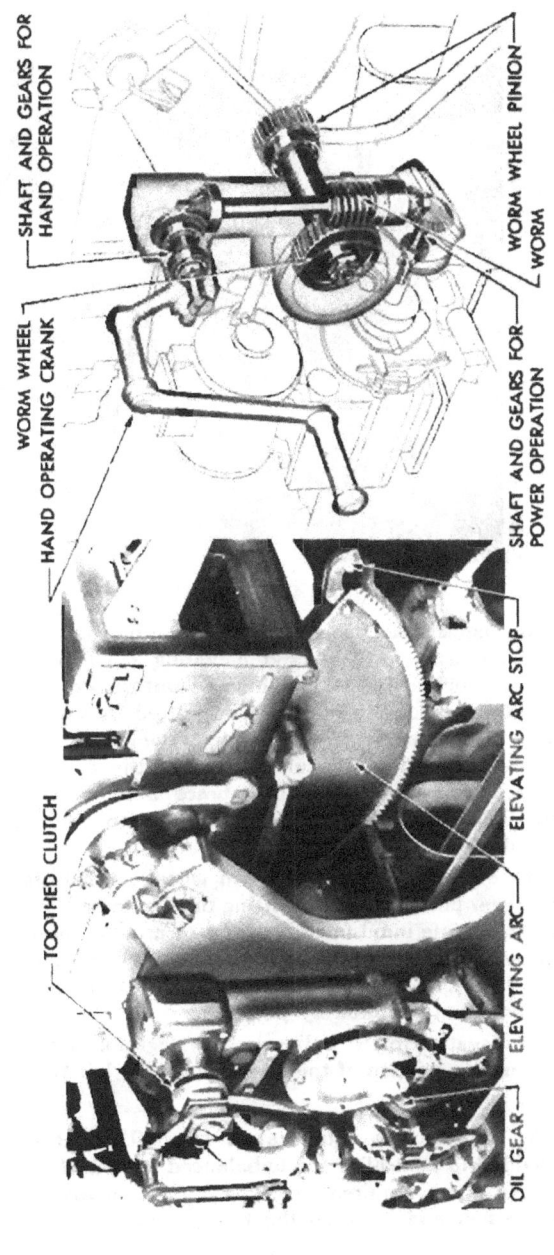

Figure 48 — Elevating Mechanism — Assembled and Phantom Views

TM 9-252
24-25

## 40-MM AUTOMATIC GUN M1 (AA) AND 40-MM ANTIAIRCRAFT GUN CARRIAGES M2 AND M2A1

Figure 49 — Equilibrators

wheel mechanism assembly. It is mounted on the outside of the left top carriage frame. The hand elevating mechanism is manually operated through a toothed clutch on a shaft turned by a double hand crank. Rotation of the shaft is transmitted through gearing to the elevating mechanism worm wheel pinion which meshes with the elevating arc fastened to the under side of the breech casing. The pinion is inclosed in the tubular brace between the top carriage frames. The elevating arc is provided with stops limiting the movement of the gun in elevation.

b. The elevating worm and worm wheel mechanism is connected to the oil gear power drive, permitting the gun to be elevated by power. The gears operate in oil-filled housings to provide means for constant lubrication. The elevating limit switch, mounted on the right gun trunnion frame, may be adjusted to limit the elevation and depression of the gun when operated by power.

c. An elevation plate, calibrated in degrees, is mounted on the breech casing a portion of the way around the left trunnion.

## 25. EQUILIBRATORS.

a. Two spring type equilibrators (fig. 49), operating as a unit, are provided to neutralize the unbalanced weight of the weapon and reduce the effort required to elevate it. The assembly is located under the gun and between the top carriage frames.

## DESCRIPTION AND FUNCTIONING OF CARRIAGE

b. Each equilibrator consists of a tubular case containing a set of three springs held in compression on an equilibrator spring rod. The two cases are held in the equilibrator trunnion bracket which is supported in trunnions in the front edges of the top carriage frames. The two equilibrator spring rods are pivoted on the equilibrator cross bar which is clamped in and projects from each side of the elevating arc.

c. The equilibrator springs are compressed as the gun is depressed, counterbalancing the muzzle-heavy weight of the tipping parts and eliminating the need for braking their descent. The energy stored in the compressed springs is released as the gun is elevated, the expanding springs exerting a pull on the elevating arc and assisting in elevating the gun.

## 26. TRAVERSING MECHANISM.

a. The traversing mechanism (fig. 50) consists of the hand traversing mechanism group assembly and the traversing mechanism group assembly.

b. Higher gear ratios in the hand traversing mechanism of the Gun Carriage M2A1 result in hand traversing that is approximately three times as fast as that of Gun Carriage M2. The principal differences between the hand traversing mechanisms of the two carriages are a larger hand traversing mechanism gear case and long and short hand traversing mechanism bevel gears of higher ratio on the Gun Carriage M2A1.

c. The hand traversing mechanism is mounted on the outside of the right top carriage frame and is manually operated through a toothed clutch on a shaft turned by a double hand crank. The rotation of this shaft is transmitted by bevel gears in the hand traversing mechanism gear case through the flexibly coupled traversing mechanism pinion shaft to the traversing gear reduction mechanism assembly to which the traversing mechanism is splined. The traversing mechanism pinion meshes with the top carriage traversing ring gear on the chassis frame base. The top carriage is rotated in azimuth by the traversing mechanism pinion being rolled around the circumference of the stationary top carriage traversing ring gear.

d. The traversing mechanism power drive extension shaft is geared to the traversing mechanism pinion shaft and connects the traversing mechanism to the azimuth oil gear, permitting the top carriage to be traversed by power. The azimuth indicator drive shaft is also geared to the traversing mechanism pinion shaft. (The azimuth indicator is described in par. 96 h.) All gears operate in oil-filled gear cases to provide constant lubrication.

## 27. POWER SYNCHRONIZING MECHANISM (SLEWING HANDLE).

a. The power synchronizing mechanism (fig. 51) provides a means

TM 9-252
27

**40-MM AUTOMATIC GUN M1 (AA) AND 40-MM ANTIAIRCRAFT GUN CARRIAGES M2 AND M2A1**

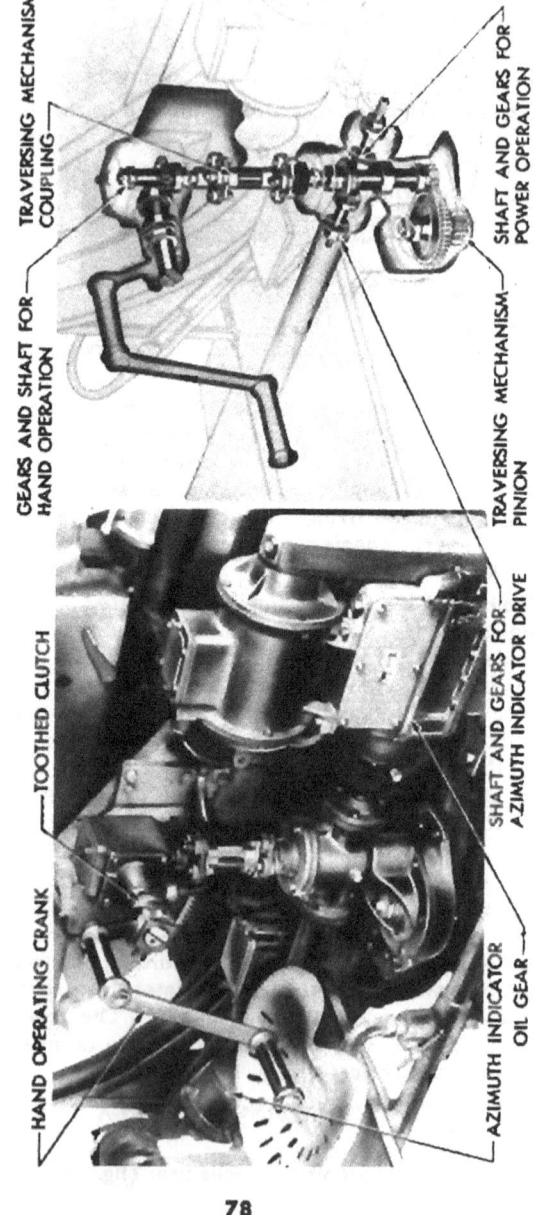

Figure 50 — Traversing Mechanism—Assembled and Phantom View

## DESCRIPTION AND FUNCTIONING OF CARRIAGE

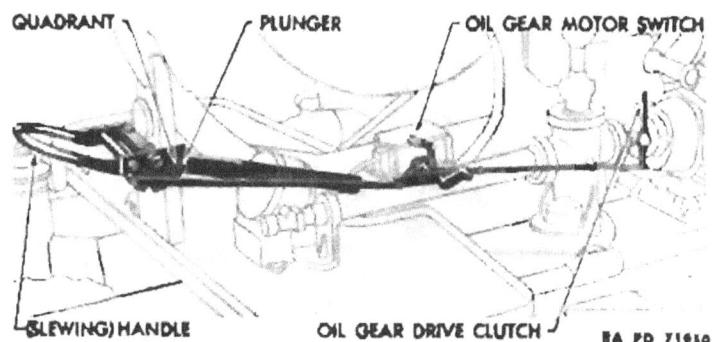

*Figure 51 — Power Synchronizing Mechanism — Phantom View*

for rapid traverse (slewing) when the weapon is being traversed manually. It also provides a means for bringing the gun into approximate alinement with the director in order that the azimuth oil gear may be engaged.

b. The power synchronizing mechanism is mounted on the rear cartridge chute body support which is mounted on the rear of the loading platform. When the power synchronizing mechanism handle is depressed, it throws the traversing oil gear motor switch after engaging the oil gear drive clutch. Raising the handle disengages the power drive and permits the top carriage to be swung around (slewed) rapidly. The handle is held in either raised or depressed position by the engagement of a spring-loaded plunger in a notched quadrant on the right side of the handle.

### 28. OIL GEAR UNITS.

a. The azimuth oil gear and the elevation oil gear supply controlled driving power for traversing and elevating the gun. They are units of the on-carriage equipment of the Remote Control System M5, designed for use with the 40-mm Gun Carriages M2 and M2A1 (AA). They are identical except for a conversion which can be made by the using arm personnel. They are fully described in section X, which also gives instructions for operation, maintenance, tests, adjustments, and instructions for conversion.

b. Each oil gear unit consists of an oil gear or hydraulic power mechanism, driven through a roller chain by a 0.6-horsepower, 115-volt, alternating-current motor which is mounted on top of the oil gear.

### 29. FIRING MECHANISM.

a. The firing mechanism on the top carriage (fig. 52) actuates the firing mechanism in the gun. This mechanism can be operated by applying pressure to either the front or rear firing pedals.

**TM 9-252**
**29-30**

### 40-MM AUTOMATIC GUN M1 (AA) AND 40-MM ANTIAIRCRAFT GUN CARRIAGES M2 AND M2A1

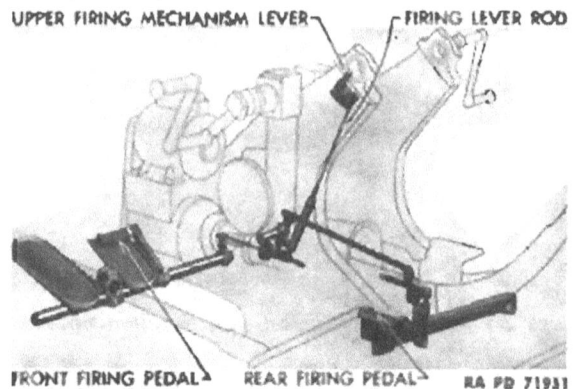

*Figure 52 — Firing Mechanism on Carriage — Phantom View*

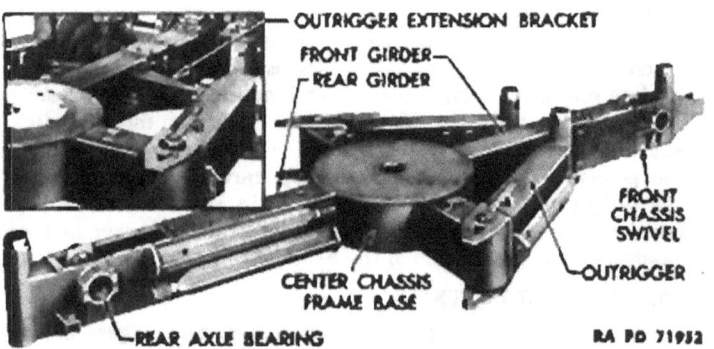

*Figure 53 — Chassis Frame With Outriggers Folded*

b. The front firing pedal is located just above the right footrest on the left side of the platform frame. It is interconnected with the rear firing pedal located just back of the seat on the same side of the gun. Pressure on either pedal is transmitted through linkage to the firing lever rod which runs up the left top carriage frame to the upper firing mechanism lever which contacts the firing crank plunger in the left gun trunnion.

c. When pressure is released, the firing pedals are returned to their normal positions by springs. Either firing pedal may be depressed without disturbing the position of the other.

### 30. CHASSIS FRAME.

a. The chassis frame (fig. 53) of the carriage is cross-shaped with

## DESCRIPTION AND FUNCTIONING OF CARRIAGE

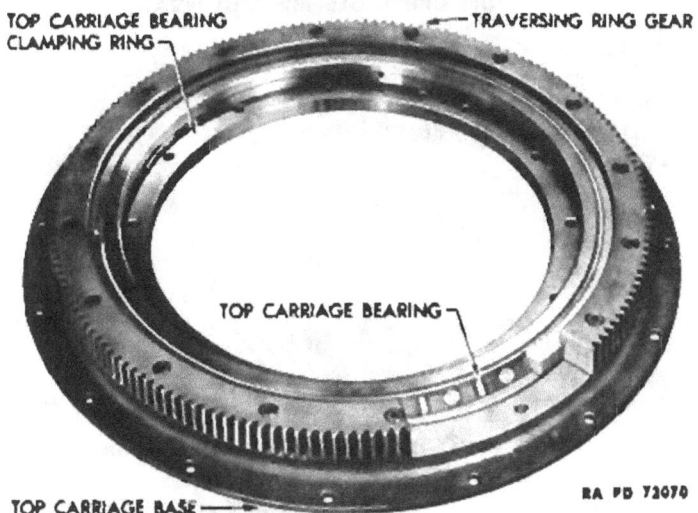

*Figure 54 — Top Carriage Roller Bearing*

a circular center chassis frame base. Long, built-up girders extend to the front and rear from the central structure. These carry the front and rear axles and suspension units, the front and rear compensating springs which aid in lowering and raising the carriage to firing and traveling positions, and the front and rear leveling jacks.

b. Short built-up arms extend to each side of the center chassis frame base. The outriggers are hinged to the side arms. Leveling jacks are mounted on ends of the outriggers. When outspread, the outriggers give sidewise stability to the carriage in firing position.

c. On carriages of late manufacture, and on those so modified, chassis frame outrigger extension brackets with double outrigger hooks (insert, fig. 53) are welded to both sides of the front chassis frame girder. When the outriggers are folded and engaged by the ends of these brackets, continuous traverse of the top carriage is permitted when the weapon is in traveling position. On unmodified carriages of early manufacture, the outriggers, when folded, are engaged by hooks welded directly to the front chassis frame girder; traverse is restricted in traveling position.

d. The circular center chassis frame base has a flat top surface. On this is mounted the top carriage ball or roller bearing which supports the top carriage, loading platform frame, and the gun with its operating equipment.

TM 9-252
30

**40-MM AUTOMATIC GUN M1 (AA) AND 40-MM ANTIAIRCRAFT GUN CARRIAGES M2 AND M2A1**

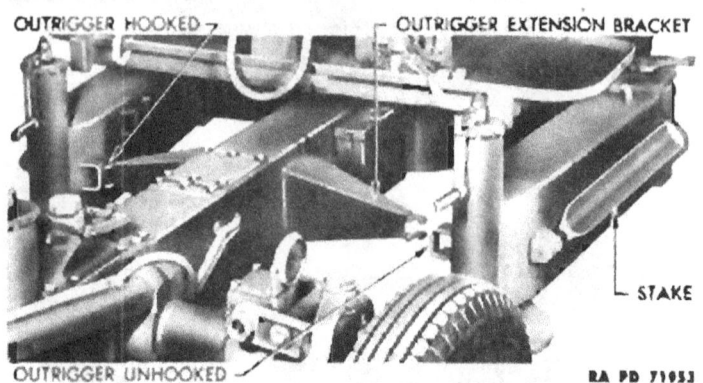

*Figure 55 — Outrigger — Folded Position*

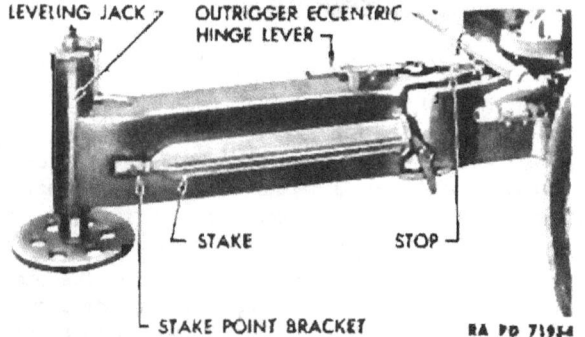

*Figure 56 — Outrigger — Extended Position*

c. **Top Carriage Ball or Roller Bearing and Traversing Gear.**

(1) The top carriage rotates on a large ball or roller bearing (fig. 54) mounted within the traversing ring gear. The upper race of the ball bearing (or the inner race of the roller bearing), which carries the top carriage, is clamped by the bearing clamping ring to the frame base of the top carriage. The lower race of the ball bearing (or the outer race of the roller bearing) is held in the traversing ring gear which is attached to the flat, circular center chassis frame base by the stationary top carriage base.

(2) The top carriage is rotated in azimuth by the traversing mechanism pinion, which is mounted on the top carriage, being rolled around the circumference of the stationary traversing ring gear.

## DESCRIPTION AND FUNCTIONING OF CARRIAGE

f. Outriggers.

(1) The outriggers are built-up box girders, similar in construction to the short transverse arms of the chassis frame to which they are hinged. The outrigger eccentric hinges permit the outriggers to be extended when the mount is in firing position (fig. 56) or folded when the mount is in traveling position (fig. 55). These hinges give inward and outward movement to the outriggers which permits them to be locked in both positions.

(2) The upper and lower plates of each outrigger extend over the top and bottom of the end of its side girder. These plates terminate in wedge-shaped toes. There are brackets on the tops and bottoms of the side members into which the toes of the plates fit when the outriggers are in firing position. Brackets on the front sides near the ends of the outriggers engage hooks to lock the outriggers in traveling position.

(3) On the top of each outrigger is a jointed outrigger eccentric hinge lever which is keyed to the eccentric hinge spindle. Each lever is equipped with a holding bracket and a stop. With the outrigger in firing position, when the lever is swung toward the rear of the carriage until its stop is contacted, the eccentric sleeves on the hinge spindle cause the outrigger to move outward. The toe of the plate is withdrawn from the bracket on the side girder. The outrigger may then be swung forward to traveling position. Here the bracket on the outrigger is caught by hooks. When the outrigger eccentric hinge lever is returned to its holding bracket, the outrigger is locked in position.

(4) On carriages of early manufacture, the brackets on the sides of the outriggers are engaged by double outrigger hooks welded to both sides of the front chassis frame girder. The result is that the outriggers are drawn so closely to the girder that traverse is restricted when the weapon is in traveling position.

(5) On carriages of late manufacture, and on those which have been modified, chassis frame outrigger extension brackets with double outrigger hooks (fig. 55) are welded to both sides of the front chassis frame girder. When the outrigger brackets are engaged in the hooks on the extension brackets, the outriggers are locked at such angles as to permit continuous traverse of the weapon when in traveling position.

(6) On carriages of early manufacture, the outrigger hinges in traveling position were fitted with welded steel outrigger hinge covers. Carriages are currently being fitted with canvas outrigger hinge covers.

g. Stakes.

(1) Four triangular-shaped chassis stakes (figs. 55 and 56) are provided to anchor the carriage to the ground in firing position. In traveling position, one of these stakes is carried on the outer side

**TM 9-252**
**30**

**40-MM AUTOMATIC GUN M1 (AA) AND 40-MM ANTIAIRCRAFT GUN CARRIAGES M2 AND M2A1**

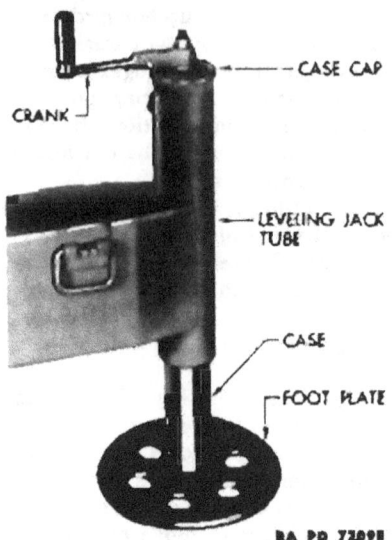

*Figure 57 — Leveling Jack With Crank Extended*

of each outrigger and two stakes are carried on the right side of the rear chassis frame girder. In firing position, the stakes are inserted in double stake brackets welded to the forward sides of the outriggers, the right side of the front girder, and the left side of the rear girder.

(2) In traveling position, the points of the stakes are fitted into cavities in the spring-loaded plungers of the stake point and the heads of the stakes are held in brackets (fig. 56).

h. **Leveling Jacks.**

(1) Leveling jack bodies are welded to the ends of the outriggers (figs. 57 and 58) and front and rear girders. These bodies house the leveling jack cases, screws, and mechanisms. The leveling jacks provide a means for leveling the mount in firing position.

(2) Each jack has a circular foot plate to afford considerable contact with the ground, and a hinged crank which is extended to operate the jack (fig. 57) or is folded downward and fitted into grooves in the case cap to lock the jack in position (fig. 58). The handle is retained in extended and folded positions by a spring-loaded plunger. The foot plate has a ball and socket mounting to compensate for uneven ground and an outer flange and a number of flanged circular openings to prevent movement on the ground.

(3) The handle of the leveling jack is keyed to the leveling jack screw (fig. 58). When the handle is turned, the leveling jack screw

## DESCRIPTION AND FUNCTIONING OF CARRIAGE

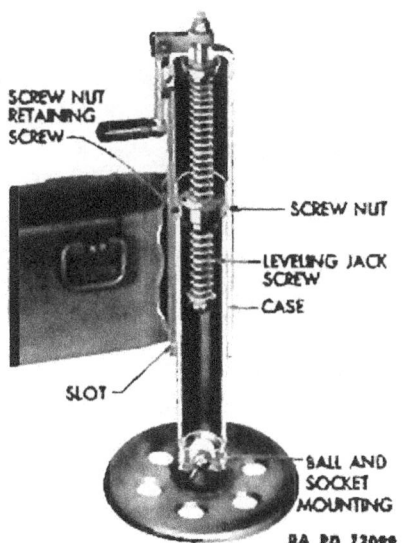

*Figure 58 — Leveling Jack — Cutaway View*

causes the leveling jack screw nut to move up or down. The screw nut is prevented from revolving with the leveling jack screw by the screw nut retaining screw which moves up and down in a slot in the leveling jack body. The leveling jack case which is attached to the screw nut, and the foot plate which is attached to the bottom of the case, are moved up and down in the leveling jack body by the action of the screw nut, lowering or lifting the weapon.

f. **Other Units Mounted on Chassis Frame.** The gun junction box (par. 96 c) is mounted on left side of the rear chassis frame girder, while the blackout light switch (par. 34 c) is mounted nearer the end on the same side of this girder. The battery container box (par. 32 d) is mounted near the center circular section on the left side of the front girder.

## 31. AXLES AND SUSPENSION.

a. In traveling position, the axles distribute the weight of the carriage to the wheels and house the spring suspension units. In addition, they may be rotated to lower and raise the carriage to firing and traveling positions. The front axle is also provided with a means of steering the carriage as it is being towed.

b. The front and rear axles are hollow tubes with bearing surfaces on which they support the carriage and seats for the parts which are attached to them. The axles support the carriage in the front chassis

**TM 9-252**

**40-MM AUTOMATIC GUN M1 (AA) AND 40-MM ANTIAIRCRAFT GUN CARRIAGES M2 AND M2A1**

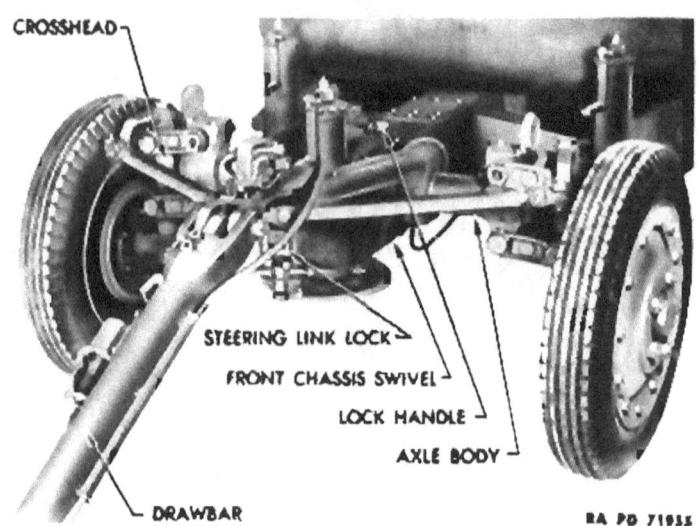

Figure 59 — Front Axle — Traveling Position

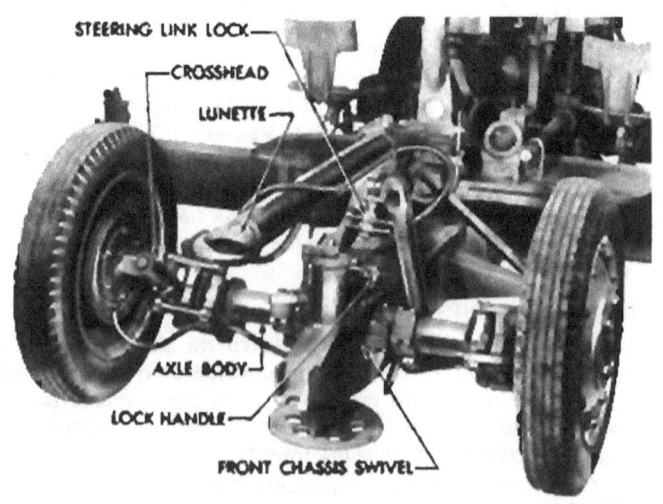

Figure 60 — Front Axle — Firing Position

TM 9-252
31

## DESCRIPTION AND FUNCTIONING OF CARRIAGE

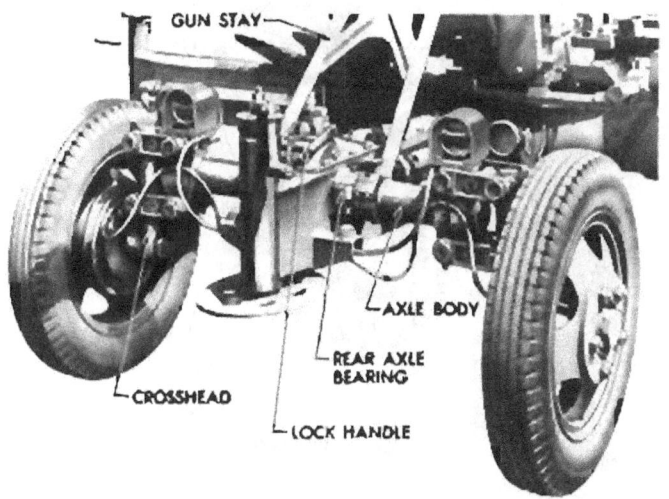

Figure 61 — Rear Axle — Traveling Position

Figure 62 — Rear Axle — Firing Position

## 40-MM AUTOMATIC GUN M1 (AA) AND 40-MM ANTIAIRCRAFT GUN CARRIAGES M2 AND M2A1

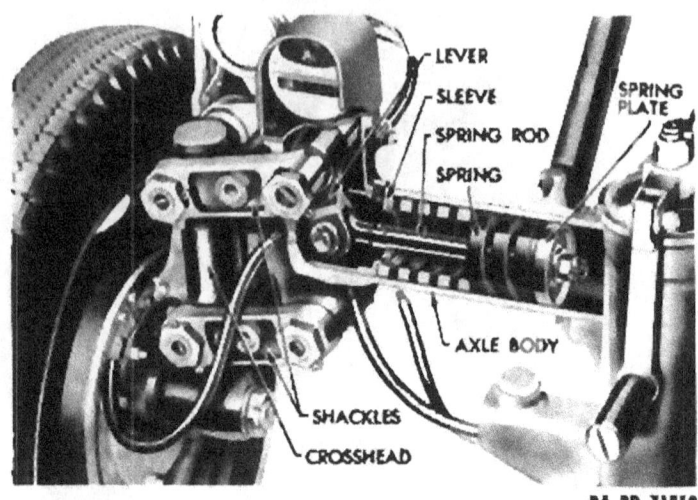

*Figure 63 — Spring Suspension — Cutaway View*

swivel and the rear axle bearings. The front and rear chassis compensating springs, arcs, and locks assist in lowering and raising the carriage to firing and traveling positions, the drawbar and gun stay being used as levers in performing these operations.

c. Spring Suspension.

(1) Each of the four wheels of the carriage has independent spring suspension, the coil spring for each wheel being housed inside its end of the axle (fig. 63). The front and rear suspension units are similar, the principal exception being that the front axle spindle shafts can be rotated in their cross heads to provide a means of steering similar to that employed in automobiles.

(2) The axles terminate in hollow, T-shaped heads. The axle cross head is attached to the top and bottom of the "T" by upper and lower shackles, permitting the cross head a limited up-and-down movement. Each front axle spindle shaft is mounted vertically in its cross head with the hub of a front axle spindle splined to the lower end of the spindle shaft. The rear axle spindles are keyed directly to hubs at the lower end of the rear axle cross heads.

(3) The axle shackle springs inside the axle bodies are compressed against the piston-like axle body sleeves by axle shackle spring rods and plates which are connected to axle shackle spring levers splined to the upper shackles. Any tendency of a wheel to move upward, as

## DESCRIPTION AND FUNCTIONING OF CARRIAGE

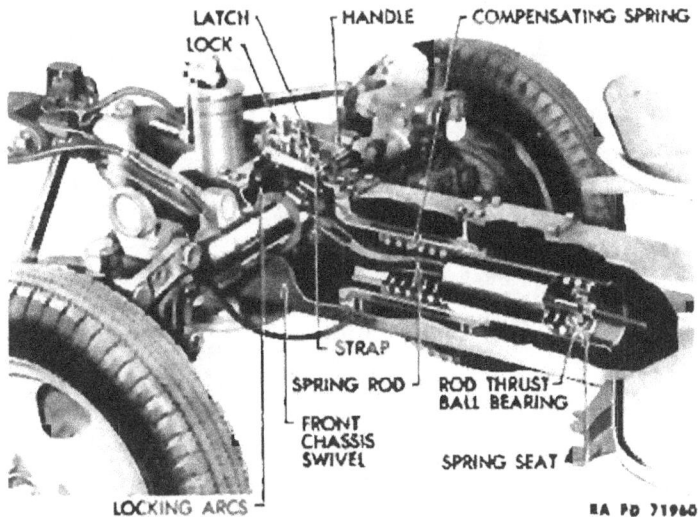

*Figure 64 — Front Chassis Compensating Spring, Arc, and Lock — Cutaway View*

in striking a bump, is counteracted by the resistance of its lever tending to move outward and compress its spring.

d. **Front Chassis Swivel.** The front chassis swivel (fig. 64) is a box-like structure with a tubular rear extension. The tubular end fits into a housing in the front end of the front girder and is secured by a collar. It not only supports the front axle but it houses the mechanism which permits the axle to be rotated about 140 degrees to allow the front part of the carriage to be raised and lowered.

e. **Front Chassis Compensating Spring, Arc, and Lock.**

(1) The function of this mechanism (fig. 64) is to assist in the lowering and raising of the carriage and to lock the axle when the carriage is in firing or traveling position. It consists of a powerful compression spring in a housing mounted in the front chassis swivel, a compression rod and seat connected to a lug on the front axle locking arc assembly, two notched locking arcs firmly keyed to the front axle, and a spring-loaded lock actuated by a locking handle.

(2) The chassis compensating spring lock handle is eccentrically mounted to cause it to raise and lower the spring-loaded lock when the handle is rotated toward and away from the gun. The lock engages notches in the arcs to lock the axle in position. The handle is automatically held in locked position by a latch.

TM 9-252
31

**40-MM AUTOMATIC GUN M1 (AA) AND 40-MM ANTIAIRCRAFT GUN CARRIAGES M2 AND M2A1**

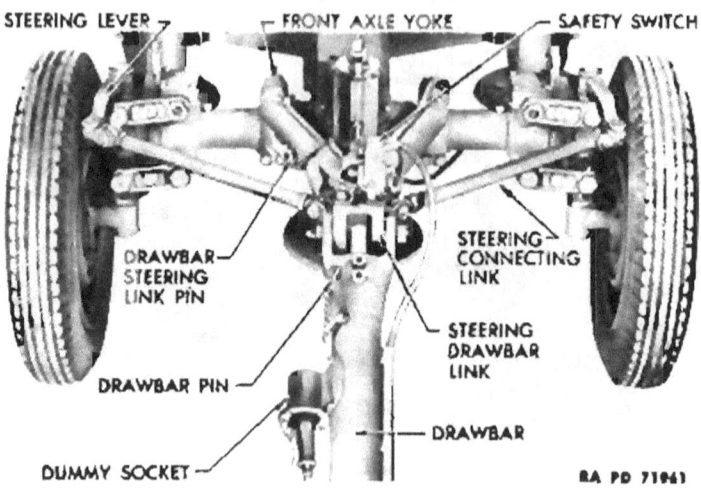

Figure 65 — Steering Mechanism

(3) Raising the lock handle lifts the lock from its notch in the arcs. When the front axle is rotated, using the draw-bar as a lever, the front end of the carriage is lowered to the ground against the compression of the compensating spring. Depressing the lock handle permits the lock to be inserted into one of five notches in both of the arcs and locks the axle in that position.

(4) When the locking handle is lifted and the drawbar is rotated away from the gun, the front axle is unlocked and rotated, and the carriage is raised with the assistance of the compressed compensating spring. Depressing the lock handle forces the lock into a notch in the arcs and the latch retains the handle, locking the carriage in traveling position.

(5) As additional assurance that the front and rear chassis compensating spring lock handles will not become unlocked inadvertently, a web strap and chassis clip have been provided for each lock handle on carriages of late manufacture and on those so modified. This strap and buckle are attached by means of the clip to one of the chassis compensating spring cover cap screws. The strap must be buckled around the bar of the lock handle to hold the handle in locked position at all times except when the handle is being operated.

f. Drawbar.

(1) The drawbar (fig. 59) is a strong tubular member which serves to connect the carriage to the prime mover, to actuate the steering mechanism of the carriage, and to provide a lever for lower-

## DESCRIPTION AND FUNCTIONING OF CARRIAGE

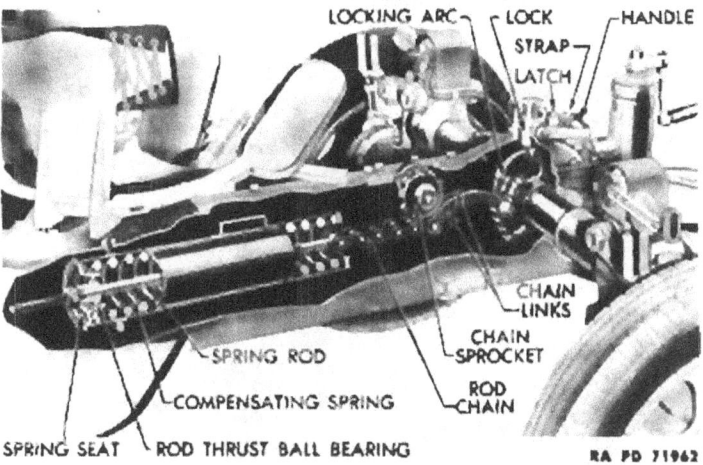

*Figure 66 — Rear Chassis Compensating Spring, Arc, and Lock — Cutaway View*

ing and raising the front part of the carriage. It is hinged to the steering drawbar link which is connected through the front axle yoke to the yoke bearings on the front axle.

(2) A lunette (fig. 60) is mounted in the forward end of the drawbar. In the earlier models, the lunette mounting was provided with rubber shock absorbing rings separated by steel washers, and was held in place in the drawbar by screws; in the later models, it is held by heavy rivets and does not have the rubber rings.

(3) To provide a rigid lever for raising and lowering the front part of the carriage, the drawbar, steering link, and front axle yoke are locked together. The drawbar steering link pin is inserted through holes near the front of the front axle yoke and near the rear of the steering link and the drawbar pin is inserted through holes near the front of the steering link and near the rear end of the drawbar.

g. Steering.

(1) Steering of the carriage is accomplished by turning the front axle spindles in unison, in the manner that an automobile is steered. The front wheels are directed to the right and left by the steering drawbar link (fig. 65) which is connected to the steering levers on the front axle spindles by the tubes and clevises of the steering connecting link assemblies.

(2) The steering drawbar link is supported by and pivots in the front end of the front axle yoke which is fastened and keyed to the front axle. The link is pivoted from side to side by the drawbar. In traveling position, the front axle yoke is held horizontally by being

## 40-MM AUTOMATIC GUN M1 (AA) AND 40-MM ANTIAIRCRAFT GUN CARRIAGES M2 AND M2A1

*Figure 67 — Gun Stay — Traveling Position*

locked to the front end of the front chassis swivel by the steering link lock (fig. 59).

h. **Rear Axle Bearing.** The rear axle bearing not only affords a mounting for the rear portion of the carriage on the rear axle, but it permits the rear axle to be rotated in lowering and raising the carriage to firing and traveling positions.

i. **Rear Chassis Compensating Spring, Arc, and Lock.**

(1) The function and operation of the rear chassis compensating spring, arc, and lock (fig. 66) are the same as those of the front chassis compensating spring, arc, and lock. The construction and action of the lock handle and lock are very much the same. There is some difference in the construction and action of the spring, arc, and other parts.

(2) The compression spring, encased in its housing, is held in the rear girder by trunnion pins and screws. The spring compression rod and seat are connected to a single locking arc by a short length of roller chain and a pair of links. The chain is carried under a ball bearing mounted sprocket when the axle is rotated to give a straight line pull on the spring compression rod. The gun stay is used as the lever for rotating the rear axle.

j. **Gun Stay.**

(1) The gun stay (fig. 67) is a tubular frame which holds the gun

## DESCRIPTION AND FUNCTIONING OF CARRIAGE

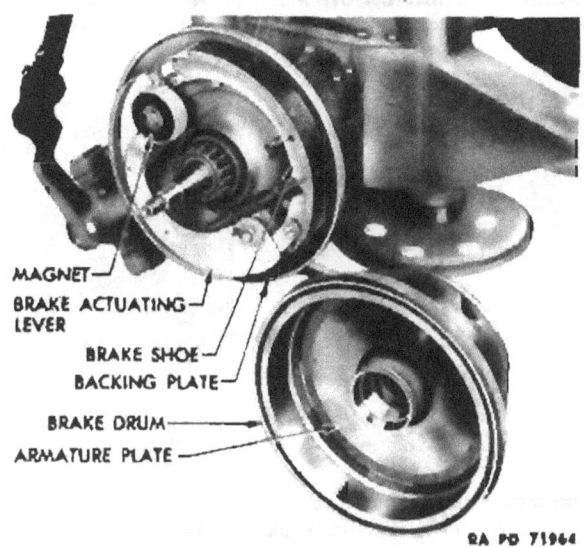

*Figure 68 — Electric Brake Mechanism and Brake Drum*

rigidly in traveling position and serves as a lever for rotating the rear axle in lowering and raising the carriage. The lower end of the gun stay pivots on the rear axle.

(2) Spring-loaded plungers at the lower ends of the gun stay engage the rear axle to form a rigid connection and convert the stay into a lever for use in placing the carriage in firing or traveling position.

(3) In traveling position, threaded plungers in the upper end of the stay are engaged in recesses in the breech casing by turning the plunger (outer) handles. The plungers are locked in place by turning the plunger locking (inner) handles.

## 32. BRAKES.

a. General. The carriage is equipped with 4-wheel electric brakes actuated by current produced on the prime mover and conducted to the carriage by a detachable cable. They are applied separately from those on the prime mover by means of a controller and a load control switch. In case of a break-in-two between the prime mover and the carriage, the brakes are applied automatically by a safety switch with electrical current supplied by a dry cell battery carried on the carriage (subpar. d below). The rear wheels are equipped with separate hand-operated mechanical brakes.

b. Electric Brake Mechanism.

(1) Each electric brake wheel unit (fig. 68) consists principally of

TM 9-252
32

**40-MM AUTOMATIC GUN M1 (AA) AND 40-MM ANTIAIRCRAFT GUN CARRIAGES M2 AND M2A1**

*Figure 69 — Safety Switch and Dummy Socket*     *Figure 70 — Battery Container Box on Front Girder*

an armature plate secured to the inside of the brake drum and rotating with it, and a magnet, brake actuating lever, and brake shoes mounted with springs to the stationary backing plate of the brake mechanism. The backing plate is mounted on a flange on the wheel spindle.

(2) The magnet is supported on the end of the brake actuating lever and is always in light contact with the armature plate. The lever is pivoted between the ends of the brake shoes and is capable of being moved backward or forward. It is returned to its normal position after it has been moved in either direction by the springs which retract the brake shoes from the drums.

(3) There is a stud on the under side of the lever at its pivoted end. This stud bears against either of two links attached to the brake shoes and presses the shoe attached to that link outwardly when the lever is moved, applying the shoe to the drum. Retracting springs withdraw the shoes from the drum when pressure upon their ends is released by the lever.

c. **Electric Brake Functioning.** When the brakes are applied, current is released which energizes the magnet (fig. 68). The magnet clings to and tends to revolve with the armature plate, causing the lever to move. The stud on the pivoted brake actuating lever presses one of the links attached to the end of a brake shoe, expanding the shoe against the brake drum and applying braking effort. The more current released, the tighter the magnet will cling to the armature plate and the harder the shoe is pressed against the drum. Application of the brakes during forward movement activates one brake shoe initially; the other brake shoe is activated initially during backward movement.

# DESCRIPTION AND FUNCTIONING OF CARRIAGE

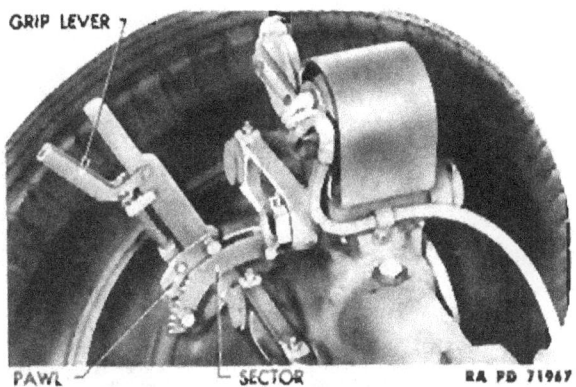

*Figure 71 — Hand Brake Lever*

d. **Safety Switch, Dummy Socket, and Battery Container.**

(1) The safety switch (fig. 69) automatically applies the electric brakes on the carriage wheels in the event of a break-in-two between the carriage and the prime mover. The switch has a lever which is normally in the "off" or backward position; it can be pulled forward to the "on" position.

(2) On carriages of earlier manufacture, the safety switch, together with the dummy socket for the jumper cable, is located on a bracket mounted on the steering drawbar link. On carriages of later manufacture and those so modified, the safety switch is mounted on the steering drawbar link but the dummy socket for the jumper cable is mounted on the side of the drawbar.

(3) The battery container box (fig. 70) is mounted near the center chassis frame base on the left side of the front chassis frame girder. It contains a dry cell battery to supply current to actuate the brakes through the safety switch in case of a break-in-two.

e. **Safety Switch Functioning.** When a break-in-two occurs, the safety chain, which connects the safety switch to the prime mover, pulls the switch lever forward. This closes the circuit from the dry cell battery to the brakes, energizes the brake magnets, and sets the brakes. The safety switch chain is provided with a spring clip which permits its being pulled from the switch lever without damage to the chain, after the lever has been pulled forward.

NOTE: The safety switch chain and the jumper cable are part of the equipment of the carriage and should be disconnected from the prime mover and retained with the carriage when the latter is uncoupled.

TM 9-252
32-33

**40-MM AUTOMATIC GUN M1 (AA) AND 40-MM ANTIAIRCRAFT
GUN CARRIAGES M2 AND M2A1**

RA PD 71965  RA PD 71969

*Figure 72 — Flat Base Rim Type Wheel*   *Figure 73 — Divided Rim Type Wheel*

f. **Hand Brakes.**

(1) The parking or mechanical hand brake mechanisms on the rear wheels are operated by hand brake levers (fig. 71) assembled to the brake backing plates. Each lever is secured in the engaged position by a spring-loaded gear sector pawl which engages the teeth of a hand brake sector. A grip lever is squeezed toward the hand brake lever to release the brakes.

(2) The hub of the brake lever is attached to the shaft of the emergency brake operating cam which contacts the emergency brake lever and link to force the brake shoe against the drum when the brake is applied.

## 33. HUBS, WHEELS, AND TIRES.

a. **Hubs.** Each wheel hub has an outer flange to which the brake drum and wheel disk are fastened by studs, and two inner chambers for the roller bearings which support the weight of the carriage on the axle spindles. The assemblies are retained on the spindles by the axle spindle washer and nut and the outer hub opening is closed by a hub cap.

b. **Wheels.** Carriages have been equipped with two types of steel disk wheels. The truck and bus wheel (flat base rim type) (fig. 72) consists of a wheel disk and flat base rim. The tire is held on the rim by a rim flange locked by a split locking side ring. The transport combat (divided rim type) (fig. 73) consists of a wheel disk and

TM 9-252
33-34

## DESCRIPTION AND FUNCTIONING OF CARRIAGE

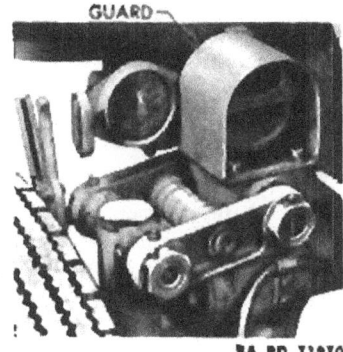
RA PD 71970

*Figure 74 — Rear Lamp Fixtures*

RA PD 71971

*Figure 75 — Blackout Light Switch*

base rim. The tire is held on the base rim by a side ring retained by 12 side ring bolt nuts. Both types of wheels are attached to the hubs by 5 wheel stud nuts. Wheel stud nuts on the right wheels have right-hand threads; those on the left wheels, left-hand threads.

c. **Tires.** Tires and tubes used on the carriage are size 6.00—20. The standard tire equipment is the 6-ply, heavy-duty, truck-bus type tire with standard heavy-duty tube and flap. Carriages of early manufacture were equipped with 6-ply, heavy-duty, truck-bus type tires and heavy bullet-resisting tubes. Some carriages were equipped with combat tires. In all cases, the pressure to be maintained in the tires is 45 pounds per square inch.

## 34 LIGHTING EQUIPMENT.

a. The light equipment (fig. 71) consists of two lamp fixtures with tail lamp guards and side and rear reflectors mounted on the tops of the rear axle cross heads, side reflectors mounted on the tops of the front axle cross heads, and a reflector mounted on the front of the cover.

b. The left lamp fixture houses the service taillight, service stop light, and a blackout taillight. The right lamp fixture houses a blackout taillight and the blackout stop light.

c. The lighting can be converted to either service or blackout by means of the blackout switch (fig. 75). It is located under the small

TM 9-252
34-36

**40-MM AUTOMATIC GUN M1 (AA) AND 40-MM ANTIAIRCRAFT GUN CARRIAGES M2 AND M2A1**

cover on the left side of the rear girder near the rear end. Turn the blackout switch shaft until the slot lines up with the letters "BO" (blackout) or "S" (service).

## Section IV

## OPERATION

### 35. INTRODUCTION.

a. This section outlines the operation of the weapon and carriage. It prescribes precautions to be taken for the protection of the personnel and materiel. Many of the operations described herein will be performed simultaneously. This section is not to be construed as a Field Manual on the service of the piece.

### 36. TO OPERATE BREECH MECHANISM.

a. To Open Breech.

CAUTION: Prior to opening the breech, place the outer safety lever in the "SAFE" position.

(1) BY HAND. Release the hand operating lever from the front latch bracket by pushing the handle inwardly toward the breech casing against spring action and lifting the handle to clear the holding catch. Lift the hand operating lever and pull it up and to the rear as far as it will go (fig. 76).

CAUTION: For safety, when the breech is open and the gun is not to be fired immediately, always place the hand operating lever

Figure 76 — Opening Breech With Hand Operating Lever

Figure 77 — Closing Breech With Outer Extractor Releasing Lever

TM 9-252
36-37

## OPERATION

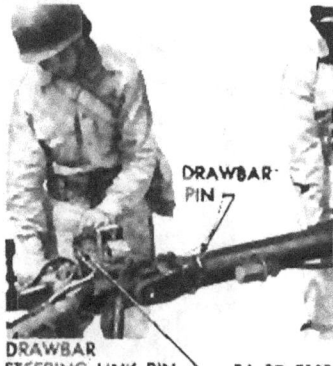

Figure 78 — Unlimbering the Weapon

Figure 79 — Locking Drawbar and Steering Yoke

in its rear latch bracket where the lever will be in vertical position and the breechblock will be held in open position. With the hand operating lever in this position, the weapon cannot be fired.

(2) AUTOMATICALLY. The functioning of the breech operating mechanism in automatic operation is described in paragraph 11 c. The hand operating lever must be released from the rear catch bracket, folded forward, and housed in the front latch bracket, for automatic loading and firing.

h. To Close Breech.

(1) BY HAND. Release the hand operating lever from the rear latch bracket and draw it to the rear to lower the breechblock slightly to take the tension off the cartridge case extractors. While doing so, press the outer extractor releasing lever rearward (fig. 77). Allow the hand operating lever to go forward easily so the breechblock will not slam shut. These operations free the breech ring outer crank to rotate, release the cartridge case extractors from the breechblock, and permit the closing spring to raise the breechblock to closed position.

(2) AUTOMATICALLY. The functioning of the breech operating mechanism in automatic operation is described in paragraph 11 c.

## 37. TO PLACE WEAPON IN FIRING POSITION.

a. Maneuver the weapon into position on approximately level ground with the prime mover.

b. Detach the safety chain from the prime mover and wrap it around the break-away safety switch. Disconnect the jumper cable and insert the plug in the dummy socket. Disconnect the drawbar lunette from the pintle of the prime mover (fig. 78). Unload the prime mover and remove it from the vicinity of the weapon.

99

# 40-MM AUTOMATIC GUN M1 (AA) AND 40-MM ANTIAIRCRAFT GUN CARRIAGES M2 AND M2A1

Figure 80 — Unlocking Steering Lock

Figure 81 — Removing Outrigger, Carriage, and Sight Covers

c. Lock the drawbar, steering link, and front axle yoke into one rigid lever. Remove the drawbar steering link pin from its bracket and insert it through the vertical holes near the front of the steering yoke and near the rear of the steering link (fig. 79). The drawbar may have to be raised and slightly moved from side to side to mate the holes. Lower the drawbar to aline holes and insert the drawbar pin through the holes near the rear of the right side of the drawbar and near the front of the right side of the steering link.

## OPERATION

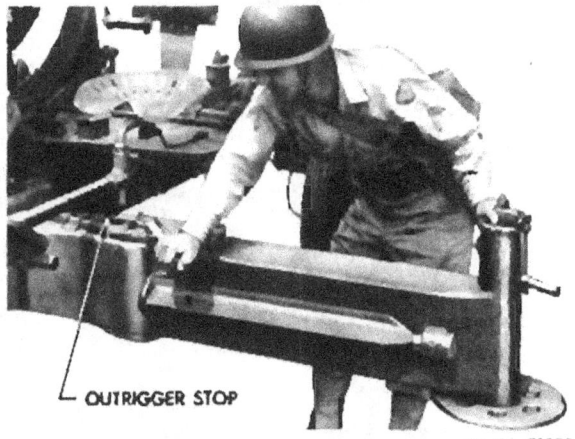

*Figure 82 — Locking Outrigger in Outward Position*

*Figure 83 — Unlocking Gun Stay and Elevating Gun*

d. Unlock the steering lock by rotating the handles (fig. 80) and swing the steering lock link forward out of its recess in the front of the front chassis swivel.

e. Remove outrigger hinge covers. Unlock the outriggers by rotating the outrigger eccentric hinge levers toward the rear of the carriage until they contact the lever stops. Swing the outriggers slightly outward to release the bottom of the carriage cover. Remove the carriage cover and the right and left sight covers (fig. 81). Check to see that the power synchronizing mechanism handle is in the disengaged or upward position.

f. Swing the outriggers to fully open position and lock them in

TM 9-252
37
**40-MM AUTOMATIC GUN M1 (AA) AND 40-MM ANTIAIRCRAFT GUN CARRIAGES M2 AND M2A1**

*Figure 84 — Unlocking Rear Chassis Compensating Spring Lock*

*Figure 85 — Unlocking Front Chassis Compensating Spring Lock*

place by returning the outrigger eccentric hinge levers to their initial positions (fig. 82).

g. Unlock the gun stay from the gun by rotating the gun stay plunger handles and locking plunger handles (fig. 83). Elevate the gun to at least 40-degree elevation.

102

**TM 9-252**
**37**

### OPERATION

*h.* Release the hand brakes if they have been set. Check the leveling jacks to make certain that they are not extended. Unbuckle the safety straps which secure the front and rear chassis compensating spring lock handles. With two men holding the drawbar and two men holding the gun stay, unlock the front and rear axles by unlatching and lifting the front and rear lock handles (figs. 84 and 85).

CAUTION: The safety straps should be fastened at all times, except when the front and rear chassis compensating spring lock handles are used to unlock the axles. The axles should never be unlocked unless there are two men to hold the gun stay and two men to hold the drawbar. The locks should be released simultaneously, and the gun stay and the drawbar must be held securely while the locks are being released.

*i.* Using the drawbar and gun stay as levers, rotate the axles until the carriage is lowered onto the foot plates of the leveling jacks (fig. 86). In performing this operation, the men at each end of the weapon must rotate the axle trees *simultaneously*. If this is not done, the balancing mechanism will not perform satisfactorily. To prevent injury to personnel, it is advisable that the carriage be lowered to the lowest position possible. If this is done, there is less likelihood of unexpected movement of the carriage when the chassis compensating spring lock handles are released in preparation for raising the carriage to traveling position. There are several positions in which the front and rear axles may be locked.

CAUTION: There will be some movement of the carriage, backward or forward, as the carriage is being lowered. Care should be taken that all members of the crew keep their feet in such positions that there is no possibility that the foot plates of front and rear leveling jacks will drop on them as the carriage is being lowered.

*j.* While the drawbar is being held vertically (fig. 87) and the gun stay is being held horizontally (fig. 88), relock the axles by returning the lock handles to their original positions. If the lock handles do not rotate easily into their locked positions, it will be necessary to raise or lower the chassis by slightly moving the gun stay or drawbar back and forth until the axles have been rotated into positions which will permit the lock handles to be returned to their locked positions. Do not force lock handles, as breakage of parts will result.

CAUTION: The drawbar and gun stay must be held securely until the lock handles have been locked. Ease the carriage into its lowered position; *do not drop it.*

*k.* Level the weapon by means of the four leveling jacks and the level assemblies mounted on the top carriage frame base (fig. 89). Do not use levels which may be mounted on the outer extremeties of the chassis frame. The carriage should be leveled as close to the ground as possible. To level the weapon on relatively even ground, place the foot plates of all the leveling jacks in contact with the

TM 9-252
37

**40-MM AUTOMATIC GUN M1 (AA) AND 40-MM ANTIAIRCRAFT GUN CARRIAGES M2 AND M2A1**

Figure 86 — Lowering Carriage to Firing Position

## OPERATION

*Figure 87 — Relocking Front Chassis Compensating Spring Lock*

*Figure 88 — Relocking Rear Chassis Compensating Spring Lock*

ground. Raise each jack three full turns of the jack handle to bring all wheels off the ground. Using the two levels mounted on the top carriage frame base and two adjacent jacks at right angles to one another, level the weapon.

l. Withdraw the drawbar pin from the holes in the right side of the drawbar and steering link, drop the steering lock link downward, and fold the drawbar downward to the horizontal position (fig. 90). Place the drawbar pin in the hole in the top of the drawbar. Unlock the gun stay from the rear axle and let it rest on its support.

CAUTION: When the weapon is in firing position, the gun stay

**TM 9-252**
**37**

**40-MM AUTOMATIC GUN M1 (AA) AND 40-MM ANTIAIRCRAFT GUN CARRIAGES M2 AND M2A1**

Figure 89 — Leveling Weapon

**TM 9-252**

## OPERATION

*Figure 90 — Folding Drawbar to Horizontal Position*

*Figure 91 — Removing Chassis Stake*     *Figure 92 — Driving Chassis Stake*

handle plungers should always be released so that the gun stay is loose on the axle.

m. Remove the four stakes by pushing their pointed ends against the spring-loaded plungers and lifting their heads from the retaining brackets (fig. 91). Insert the stakes in the double brackets provided for them on the chassis and outriggers (fig. 92). Using the maul, drive the stakes firmly into the ground as far as they will go.

NOTE: If the front stake cannot be driven into the ground to the point where it will not interfere with the traversing of the weapon, remove it entirely. In driving the front stake, care must be taken not to damage the break-away safety switch. This may be done by using a drift, after the stake has been started.

## 40-MM AUTOMATIC GUN M1 (AA) AND 40-MM ANTIAIRCRAFT GUN CARRIAGES M2 AND M2A1

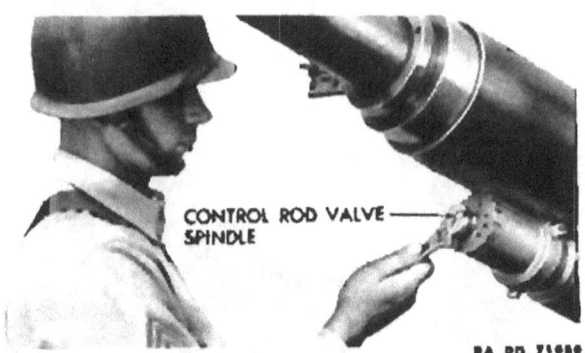

*Figure 93 — Adjusting Control Rod Valve*

n. After the stakes have been driven into the ground, check the level of the weapon in the original leveling position and for full traverse at 90-degree intervals. Make any necessary corrections. If the weapon cannot be leveled for 360-degree traverse with the level assemblies mounted on the base of the top carriage, check the level of the weapon with a gunner's quadrant. Place the gunner's quadrant on the machined portion of the breech casing and check with the gun in four positions with 90 degrees of traverse between each reading. If the levels are found to be out of alinement, adjust the faulty level as soon as conditions will permit.

o. The double crank handles of the elevating and traversing hand drive mechanisms should be attached at all times, but they should remain in the disengaged position unless the weapon is being manned for direct fire sight control.

### 38. TO CHECK LIQUID IN RECOIL CYLINDER.

a. Elevate the gun to 25 degrees. Remove the filling plug from the recoil cylinder and note whether the liquid is up to the level of the filling plug. If the cylinder is not filled to the proper level, follow the procedure for filling outlined in paragraph 66 c.

b. To Adjust Recoil Cylinder Control Rod Valve. Remove recoil cylinder control rod valve spindle adjustment screw and plate. Turn the squared end of the control rod valve spindle (fig. 93) to close the control rod valve fully; then back off the spindle one-third turn. Lock the spindle with the adjustment screw and plate. A further slight adjustment may be necessary after the gun is fired.

### 39. TO TRAVERSE.

a. By Hand. The weapon is traversed by turning the double

## OPERATION

*Figure 94 — Traversing Weapon*

hand crank on the right side of the gun (fig. 94). A seat and footrests are provided for the operator who moves in azimuth with the gun. Turning the crank clockwise traverses the gun to the left. Traverse is continuous. On the Gun Carriage M2, one turn of the crank moves the gun approximately 6 degrees in azimuth. Approximately 58 turns of the crank are required to traverse the top carriage 360 degrees. On the Gun Carriage M2A1, one turn of the crank moves the gun approximately 17 degrees in azimuth. Approximately 21 turns of the crank are required to traverse the top carriage 360 degrees.

b. **Rapid Traverse.** Pulling up on the handle of the power synchronizing mechanism (slewing handle) disengages the traversing power drive and permits the top carriage to be swung around (slewed) rapidly (fig. 95). This lever must be depressed to engage the traversing power drive and permit the top carriage to be moved in azimuth by the director.

c. **By Power.** To permit the weapon to be traversed by the director the hand drive traversing mechanism must be disengaged by pulling the double hand crank out of mesh (away from the gun), and depressing the handle of the traversing power drive (slewing handle) to engage the traversing power drive.

**40. TO ELEVATE.**

a. **By Hand.** The gun is elevated by turning the double hand

TM 9-252
40

## 40-MM AUTOMATIC GUN M1 (AA) AND 40-MM ANTIAIRCRAFT GUN CARRIAGES M2 AND M2A1

Figure 95 — Rapid Traverse — Slewing Gun

Figure 96 — Elevating Weapon

crank on the left side of the gun (fig. 96). A seat, footrests, and firing pedal are provided for the operator who moves in azimuth with the gun. Turning the crank clockwise elevates the gun. The maximum elevation is 90 degrees. The maximum depression is minus degrees. One turn of the crank elevates the gun approximately degrees.

110

## OPERATION

*Figure 97 — Removing Flash Hider Cover and Automatic Loader Hood and Shield*

b. **By Power.** To permit the gun to be elevated by the director, the hand drive elevating mechanism must be disengaged by pulling the double hand crank out of mesh; then the elevation oil gear clutch must be shifted to "on" position, if it is not already in that position, and finally, the elevating limit switch must be placed in "on" position. Maximum power elevation and depression are limited by the setting of the limit switch and are approximately 85 degrees and zero degree, respectively.

## 41. PRIOR TO FIRING.

a. If conditions permit, check the level of the liquid in the recoil cylinder (par. 38 a). Check the adjustment of the control rod valve (par. 38 b). Inspect the recoil cylinder for oil leakage.

CAUTION: Close and lock the top cover before firing.

b. Make certain that the breech casing top cover is closed and locked securely. Serious damage will be done by firing the gun with the top cover open. Not only is the barrel assembly unlocked and free to turn in the breech ring when the top cover is open, but the breech ring is locked in position in the breech casing by the breech ring barrel catch control arm. When the gun is fired with the control arm in this position, the arm is sheared off and jammed between the breech ring and breech casing.

c. Make certain that the director, carriage stakes, or any other part of the materiel does not interfere with the traversing of the weapon.

d. Just prior to loading and firing the gun, remove the flash hider cover and the automatic loader hood and shield (fig. 97). These covers should be in place as much of the time as possible to protect the bore and loader from dirt and other foreign matter. Check to make

TM 9-252
41-43

## 40-MM AUTOMATIC GUN M1 (AA) AND 40-MM ANTIAIRCRAFT GUN CARRIAGES M2 AND M2A1

*Figure 98 — Inserting Clip Into Automatic Loader*

sure that the outer safety lever is in the "SAFE" position. Inspect the bore for foreign matter and clean if necessary. If conditions permit, wipe or clean the bore to remove the coating applied after the previous firing.

### 42. TO LOAD.

a. Pull the hand operating lever at the left side of the breech casing to the rearmost position and engage it in the rear latch bracket. This operation opens the breech, cocks the firing mechanism, pulls the rammer shoe to the rear where it is engaged, and releases the feed rollers to permit them to be rotated.

b. Place a cartridge on the loading tray by pushing a clip of ammunition down the guides of the loader (fig. 98) until the feed rollers are rotated and a cartridge drops on the loading tray.

c. If the gun is not to be fired immediately, the hand operating lever should remain in the rear latch bracket to make the gun safe against accidental firing.

d. To Load Ammunition in Clips. Press the spring-loaded pin on the side of the clip (fig. 99). Insert the rim of the cartridge in the proper one of the four concave depressions of the clip. Release the pin to permit the hook to clamp the cartridge in place in the clip.

### 43. TO FIRE.

a. Immediately before firing is to take place, the hand operating lever must be folded forward and placed in the front latch bracket. The outer safety lever must also be in "SINGLE FIRE" or "AUTO FIRE" position before firing can occur.

112

TM 9-252
43

## OPERATION

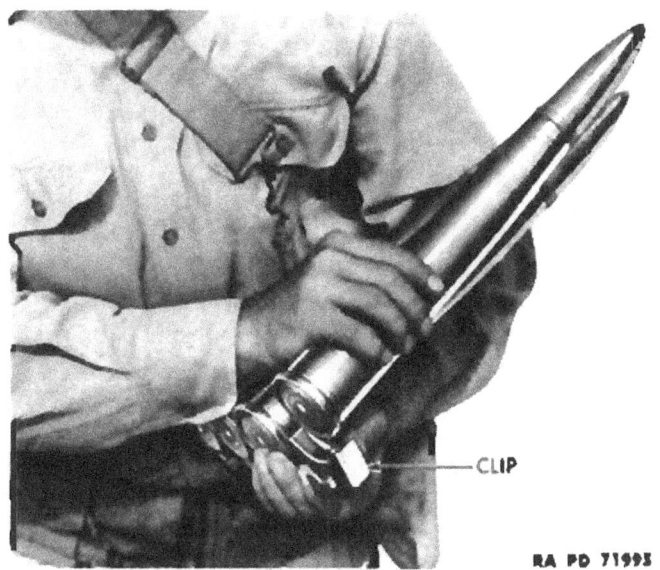

*Figure 99 — Loading Cartridge in Clip*

*Figure 100 — Firing Weapon With Rear Firing Pedal*

TM 9-252
43-44

**40-MM AUTOMATIC GUN M1 (AA) AND 40-MM ANTIAIRCRAFT GUN CARRIAGES M2 AND M2A1**

Figure 101 — Recoil Indicator Set to Measure Length of Recoil

Figure 102 — Loading Tray Striking Indicator To Measure Recoil

b. The firing mechanism is actuated by either of two controls. The front firing pedal is located above the elevation operator's right footrest on the left side of the weapon. The rear firing pedal is located near the rear of the loading platform on the left side of the breech (fig. 100).

c. To fire the gun, either pedal is depressed. With the outer safety lever set at "SINGLE FIRE," the pedal must be depressed to fire each round separately. With this lever set at "AUTO FIRE," the gun will continue firing as long as the pedal is depressed and there are more than two rounds in the automatic loader.

**44. OBSERVATION DURING FIRING.**

a. Observe the movement of the weapon in recoil and counterrecoil. Movement should be smooth and the end of movement in both directions should be without shock. During action, the operation of the recoil system should be watched carefully and necessary corrective measures must be taken to insure that the weapon will not be forced out of action due to malfunction of the recoil system. Examine the recoil cylinder at intervals during fire to be sure that there is no leakage of liquid.

b. Measure Length of Recoil.

(1) Measure the length of recoil for the first round and at intervals during firing, when practicable. The normal length of recoil is from 7.4 to 8.3 inches. The most desirable length is 7.87 inches. Excessive recoil will cause damage to the mechanism. Short recoils will prevent the effective working of the automatic loader.

(2) To measure the length of recoil, press the lower arm of the

## OPERATION

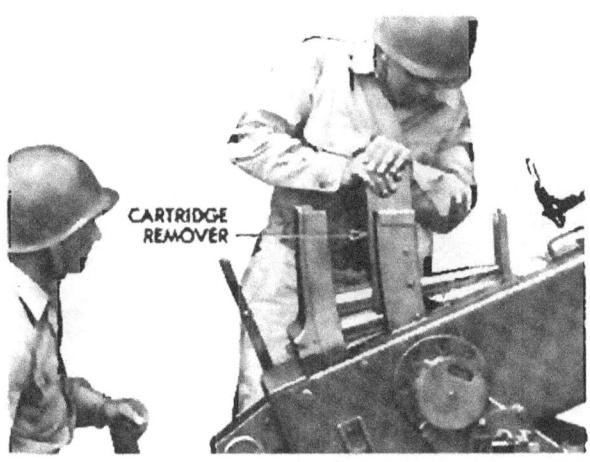

*Figure 103 — Inserting Cartridge Remover*

recoil indicator pointer (fig. 101) forward, causing the pointer to register "60" (6 in.) on the recoil indicator scale plate. When the piece is fired, the rear end of the loading tray will strike the lower arm of the indicator pointer in recoil (fig. 102). The pointer will indicate the farthest backward movement of the tray. Read the measurement on the scale plate which is calibrated in tenths of inches from 6 to 9 inches.

(3) If the length of recoil does not fall within the limits designated, take corrective measures at once (par. 53).

c. Control Speed of Counterrecoil.

(1) The counterrecoil of the gun has an important bearing on the steadiness and service of the weapon. A violent counterrecoil will cause undue shock to the carriage; a slow counterrecoil will result in slow firing. Both are to be avoided.

(2) The squared head of the control rod valve spindle, which protrudes from the front end of the recoil cylinder, provides a means for adjusting the speed of counterrecoil. The setting of this spindle should be checked before firing is commenced, to see that the control rod valve is opened one-third turn of the spindle. If the speed of counterrecoil is unsatisfactory, the setting of this spindle should be adjusted to correct the condition (par. 38 b). The valve should be closed slightly if the speed of counterrecoil is too great; it should be opened slightly if the speed is too slow.

(3) It will be noted that the gun can counterrecoil more easily when it is at a lower elevation because the effect of gravity on the gun is lessened. Watch for too severe counterrecoil when the gun is fired at lower elevations. Report any irregularity of operation at once.

**40-MM AUTOMATIC GUN M1 (AA) AND 40-MM ANTIAIRCRAFT GUN CARRIAGES M2 AND M2A1**

Figure 104 — Raising Cartridges Into Cartridge Remover

Figure 105 — Removing Cartridges From Automatic Loader

## 45. TO UNLOAD.

a. Before attempting to unload the gun, place the outer safety lever in "SAFE" position and elevate the gun to about 30 degrees. Pull the hand operating lever all the way back past the rear latch bracket. A man at the rear of the gun should reach in through the opening in the rear cover and catch the round as it slides out of the firing chamber as the lever is being pulled backward.

b. The rounds in the automatic loader are removed with the cartridge remover, which depresses the feed and stop pawls. Insert the remover into the loader (fig. 104) until the top of the remover is flush with the top of the feed and stop pawl holder frame.

CAUTION: Do not exert excessive force on the cartridge remover, for damage to the feed rollers will result.

c. With the aid of the shell pusher, manipulated through the opening in the rear cover, work the cartridges up into the remover and lift the cartridges and remover through the top of the loader (fig. 105).

d. With the hand operating lever still in the extreme rearward position, trip the extractor releasing lever at the right front of the breech casing. The breechblock is then free to be raised by the action of the breechblock closing spring, the movement being controlled by moving the hand operating lever to its forward (horizontal) position.

e. Place the feed control thumb lever to the right-hand position. Step on the firing pedal to release the rammer. The gun is now unloaded and uncocked.

## OPERATION

*Figure 106 — Withdrawing Carriage Stakes From Ground*

f. **Cartridge Case Removal With Hand Cartridge Extractor.**

(1) If the cartridge case sticks in the chamber and cannot be removed in the normal manner, it should be pried out with the hand cartridge extractor.

(2) Open the breech. Open the top cover. Insert the hand cartridge extractor through the upper opening of the breech casing into the aperture in the breech end of the tube. Engaging the rim of the cartridge case with the extractor, carefully pry the case out of the chamber. Check the condition of the extractors in the breech ring to determine the cause of the faulty extraction. Also check the firing chamber for pitting or excessive foreign material.

g. **Cartridge Case Removal With Hand Shell Ejector.**

(1) If the cartridge case cannot be removed in the normal manner or with the hand cartridge extractor, or if the case separates from the projectile, the round or the projectile should be removed with the hand shell ejector.

(2) Place the weapon at zero elevation. Open the breech. Assemble the cleaning staff M14. Screw the hand shell ejector onto the end of the assembled cleaning staff. Insert the ejector into the muzzle of the gun and push the round back onto the loading tray. Make a thorough check of the firing chamber and extractors to determine the cause of the malfunction.

**46. TO PLACE IN TRAVELING POSITION.**

a. Withdraw the carriage stakes from the ground by inserting the end of a crowbar in successive holes in the rear of the stake, prying the stake from the ground (fig. 106).

### 40-MM AUTOMATIC GUN M1 (AA) AND 40-MM ANTIAIRCRAFT GUN CARRIAGES M2 AND M2A1

b. Raise the drawbar to vertical position and insert the drawbar pin in the holes in the drawbar and steering link. Lock the gun stay to the rear axle by means of the gun stay handle plungers.

c. Raise the leveling jacks to place all four wheels of the carriage in contact with the ground. If due to rough or sloping terrain holes have been dug beneath the wheels for emplacing, these holes must be refilled if necessary to insure that the wheels are actually on the ground. After the leveling jacks have been raised to traveling position, check to see that the jack handles are in locking notches which face directly away from the gun; otherwise, difficulty will be experienced in rotating the gun stay and drawbar when the carriage is being raised.

CAUTION: Under no circumstances will the locking handles which lock the front and rear axles be unlocked unless the carriage wheels are in solid contact with the ground. Failure to observe this precaution is likely to result in serious injury to battery personnel.

NOTE: The effort of raising the carriage from firing to traveling position is aided by the strong chassis compensating springs in the front and rear girders. They operate by pushing the wheels against the ground. If the lock handles are rotated before the wheels are lowered to the ground, the force of the compensating springs is released to rotate the axles violently. The result will be that personnel in the way of the drawbar or gun stay will be injured.

d. With the carriage wheels on the ground, and the gun stay and drawbar held tightly to resist any unexpected movement, unlock the front and rear axles by rotating the lock handles.

e. Using the drawbar and gun stay as levers, rotate the axles to raise the carriage to traveling position. Raise both ends of the carriage at the same time, or raise the rear end first. Raise the carriage as gradually as circumstances permit to prevent it from "bouncing" into traveling position.

CAUTION: Do not stand directly in the path of the drawbar or gun stay until the carriage has been securely locked in traveling position. Keep feet from under the foot plates of the front and rear leveling jacks.

f. Lock the axles in traveling position by rotating the lock handles. Lock the steering link securing clamp. Depress the gun, unlock the gun stay from the rear axle, and lock the gun stay to the breech casing. Disengage the traversing and elevating double hand cranks by pulling them outwardly from the gun; leave them attached to the weapon. Remove the drawbar steering link pin and place it in its bracket. Remove the drawbar pin and place it in the hole in the top of the drawbar.

## OPERATION

*Figure 107 — Weapon in Traveling Position — Covers in Place*

g. After replacing the carriage cover and right and left sight covers, unlock, close, and lock the outriggers. Replace the outrigger hinge covers (fig. 107).

h. Back up the prime mover until the drawbar lunette can be placed over the pintle of the prime mover. Connect the jumper cable and safety chain. Make sure that the safety chain is not connected in such a manner as to cause the brakes to function on a sharp turn; leave sufficient slack to prevent this from happening.

## 47. BRAKE OPERATION.

a. When applying the brakes for an ordinary stop, the handle of the controller on the prime mover should be advanced gradually for a light brake operation. Heavy brake application should be reserved for emergency stops and should not be employed in ordinary brake service.

b. The load control on the controller is provided to permit the driver to regulate the braking power and prevent skidding regardless of load and road conditions. For slippery roads, the control should be set on "LIGHT."

c. When applying the brakes for a slowdown or stop, always apply the brakes on the load before applying the brakes on the prime mover.

d. When the speed of the prime mover is above the idling speed of the engine, leave the engine in gear and close the throttle. This will permit the engine to function as a brake. When the speed of the vehicle corresponds to the idling speed of the engine, throw out clutch and gradually apply brake.

e. When brakes are new, several applications must be made before maximum efficiency is obtained, because the magnet must wear grooves in the armature to insure proper contact.

TM 9-252
47-48

**40-MM AUTOMATIC GUN M1 (AA) AND 40-MM ANTIAIRCRAFT GUN CARRIAGES M2 AND M2A1**

CAUTION: To avoid injury to the personnel, to insure safe road transportation, and to prevent "jackknifing" of the load, the driver should have the load under control at all times by avoiding any slack between the load and the prime mover. On down grades, curves, and rough or slippery roads, the speed should be held to approximately 10 miles per hour.

Section V

## MALFUNCTIONS AND CORRECTIONS

### 48. MISFIRE.

a. In the event a round does not fire when the firing pin hits the primer, the following sequence of operations should be adhered to in replacing the unfired round with a fresh round:

(1) Recock immediately without opening the breech and refire. If the round does not function upon the second attempt, wait 30 seconds from the time the original misfire occurred. Then remove the round.

NOTE: To recock the 40-mm Automatic Gun M1 (AA) without opening the breech, lift the hand operating lever slightly until the inner cocking lever is engaged by the check plunger; then return the hand operating lever to the normal firing position. The former operation will permit the breechblock to drop about ¼ inch and cock the percussion mechanism; the latter operation will return the breechblock to the fully closed position and release the firing pin to strike the primer. *Do not perform these operations with the top cover open.*

(2) If it is impossible to remove the round in the normal manner within 45 seconds after the original misfire, and if the gun is hot, play water on the barrel. All personnel must stand clear of the gun. When the gun is cool, remove the round. If water is not available, stand clear of the gun until it is cool; then remove the round.

b. Removal of the round in the normal manner consists of elevating the gun to about 30 degrees and pulling the hand operating lever completely back past the rear latch bracket, a man being placed at the breech to catch the round as it slides out while the lever is being pulled backward.

c. If the round cannot be removed in the normal manner, due to the round sticking in the tube or the case separating from the projectile, remove the round with the hand cartridge extractor (par. 45 f), or with the hand shell ejector mounted on the cleaning staff M14 (par. 45 g).

## MALFUNCTIONS AND CORRECTIONS

NOTE: The possibility of a hangfire of more than 20 seconds after an attempt to fire is very remote in guns using fixed ammunition. The possibility of the propellent or the high-explosive filler being fired by the heat absorbed from a hot gun barrel increases with the length of time the round is in the gun. The safest time to remove a misfire is between 30 and 45 seconds after its occurrence.

### 49. BREECHBLOCK WILL NOT CLOSE.

a. In addition to the types of malfunction wherein the firing pin is released but fails to detonate the round, or the breechblock remains in open position due to mechanical obstruction, malfunctions have occurred wherein the breechblock is apparently closed, or very nearly so, and the firing pin is not released. Malfunctions of this type usually are the result of the following:
  (1) Improper lubrication of the breech mechanism.
  (2) Sand or burs in the breech mechanism.
  (3) Deformation of the cartridge case.

b. Normally, such malfunctions may be corrected by proper lubrication of the breech mechanism, removal of sand and burs in the breech mechanism, or by the use of the hand cartridge extractor or the hand shell ejector to remove the cartridge case; however, instances have occurred in which a round of ammunition has become stuck in the firing chamber due to sand in the chamber or deformation of the cartridge case and at the same time the breechblock failed to close.

c. A malfunction of this type readily can be identified by the fact that the breechblock appears to be closed (or very nearly so) and cannot be opened by pulling the hand operating lever to the rear. This, of course, is due to the fact that the cartridge case jammed in the chamber will not permit the cartridge extractors to move to the rear, thus preventing the breechblock from descending.

d. In correcting malfunctions of this nature, particular care must be exercised to insure that the breechblock will not accidentally close and fire the round while the malfunction is being corrected. Failure to take necessary precautions will result in injury to personnel and damage to materiel. To clear malfunctions of this type, the following steps should be taken by the using arm personnel after the second attempt to fire has been made and the gun has cooled:
  (1) Insure that the outer safety lever is on "SAFE," that the gun is pointed to a safe field of fire, and that the weapon is elevated approximately 30 degrees.
  (2) Pull the hand operating lever as far back as it will go and maintain a steady tension or pull on the lever tending to pull the breechblock down.

## 40-MM AUTOMATIC GUN M1 (AA) AND 40-MM ANTIAIRCRAFT GUN CARRIAGES M2 AND M2A1

(3) While tension is being maintained on the hand operating lever, remove the top cover and use a wood block and maul to drive the breechblock down into cocked position.

(4) Catch the extracted round as it slides out of the chamber. If the cartridge does not freely slide out of the chamber, it may be removed either with the hand cartridge extractor or the hand shell ejector.

c. In the event that the malfunction cannot be corrected by this procedure, notify ordnance maintenance personnel.

## 50. FEED CONTROL THUMB LEVER TO LEFT AND GUN FAILS TO FIRE.

a. Rammer Shoe to Rear and Round on Loading Tray.

| Cause | Correction |
|---|---|
| Insufficient ammunition in automatic loader. | Insert loaded clip to continue firing. |
| Broken rammer shoe plunger spring. | Notify ordnance maintenance personnel. |
| Broken rammer spring or springs. | Notify ordnance maintenance personnel. |
| Firing mechanism not properly adjusted. | Make required adjustment (par. 67 f). |
| Rammer check levers or catch lever out of adjustment and will not release rammer shoe. | Notify ordnance maintenance personnel. |

b. Rammer Shoe Forward and No Round on Loading Tray.

| Cause | Correction |
|---|---|
| Jammed feed pawls. | Unload and release feed pawls. |
| Rounds not properly loaded in automatic loader. | Unload and load properly. |
| Loading tray jammed on feed rod roller or rollers. | Notify ordnance maintenance personnel. |

c. Rammer Shoe Forward and Round on Loading Tray.

| Cause | Correction |
|---|---|
| Rammer check levers or catch lever worn, damaged, or out of adjustment. | Notify ordnance maintenance personnel. |
| Gun breech side cover unlocked during firing. Breechblock did not open. | Carefully remove round that has been slammed against breechblock. Remove breech mechanism to inspect for possible damage. Inspect side cover latch. Request ordnance maintenance personnel to replace worn or damaged parts. |

## MALFUNCTIONS AND CORRECTIONS

### 51. FEED CONTROL THUMB LEVER TO RIGHT AND GUN FAILS TO FIRE.

**a.** No Round on Loading Tray, Rammer Shoe Forward, Breech Open, and One Round in Feed Pawls.

| Cause | Correction |
|---|---|
| Insufficient ammunition in automatic loader. | Unload round if firing is to be discontinued. Reload if firing is to be continued. |
| Feed control thumb lever linkage incorrectly assembled, or out of adjustment. | Notify ordnance maintenance personnel. |

**b. Misfeed.** No Round Brought Down on Loading Tray.

| Cause | Correction |
|---|---|
| Clips loaded with bottom round not parallel to top of loader. | Push down on rounds; if resistance is met, remove all rounds from loader and inspect for irregularities in ammunition or clip. |
| Improper assembly of feed rods. | Notify ordnance maintenance personnel. |

**c.** Rammer Shoe Forward, Round in Chamber, and Breechblock Partially Closed (par. 49).

| Cause | Correction |
|---|---|
| Excessive grease in firing chamber. | Unload, clean chamber thoroughly, and inspect. |
| Burs or scoring on parts of breech mechanism or breech ring. | Notify ordnance maintenance personnel. |
| Breech ring outer crankshaft bent. | Disassemble breech mechanism and replace damaged parts. |
| Loose projectile. | Remove round with hand shell ejector. |
| Breech ring closing spring broken. | Remove closing spring case assembly and replace spring. |

**d.** Rammer Shoe Forward, Round in Chamber, and Breechblock Open.

| Cause | Correction |
|---|---|
| Breech ring closing spring broken. | Remove closing spring case assembly and replace spring. |

**e.** Rammer Shoe Forward, No Round in Chamber, and Breechblock Open.

| Cause | Correction |
|---|---|
| Broken feed roller catch head, catch head arm, or catch head spring. | Notify ordnance maintenance personnel. |

**f.** Rammer Shoe Forward, Round on Loading Tray, and Breechblock Open.

| Cause | Correction |
|---|---|
| Rammer check levers or catch lever worn, broken, or out of adjustment. | Notify ordnance maintenance personnel. |

TM 9-252
51-52

## 40-MM AUTOMATIC GUN M1 (AA) AND 40-MM ANTIAIRCRAFT GUN CARRIAGES M2 AND M2A1

| Cause | Correction |
|---|---|
| Gun breech side cover unlocked during firing. Breechblock did not open. | Carefully remove round that has been slammed against breechblock. Remove breech mechanism to inspect for possible damage. Inspect side cover latch. Request ordnance maintenance personnel to replace worn or broken parts. |

*g. Rammer Shoe Forward, Round in Chamber, and Breechblock Closed.*

| | |
|---|---|
| Broken firing pin. | Remove breechblock and disassemble percussion mechanism. Replace worn or broken parts. Check protrusion of firing pin with gage before installing breechblock. |
| Firing pin protrusion insufficient. | Replace firing pin. Check firing pin protrusion with gage before installing breechblock. |
| Firing pin incorrectly assembled. | Assemble correctly. |

## 52. OTHER MALFUNCTIONS OF GUN.

*a. Cartridge Case Partially Ejected, or Failure to Eject.*

| | |
|---|---|
| Broken rammer spring. (Rammer levers retard movement of spent cartridge case.) | Notify ordnance maintenance personnel. |

*b. Failure to Single Fire.*

| | |
|---|---|
| Firing mechanism not properly adjusted. | Make required adjustment (par. 67 f). |
| Rammer will not release. Broken or missing rammer lever plunger spring. | Notify ordnance maintenance personnel. |
| Firing lever pawl rusty or damaged. | Notify ordnance maintenance personnel. |
| Firing lever pawl plunger spring broken or missing. | Notify ordnance maintenance personnel. |

*c. Hand Operating Lever Will Not Cock Rammer Shoe.*

| | |
|---|---|
| Firing rod linkage incorrect. Hand operating lever shaft striking front nose of firing lever. | Notify ordnance maintenance personnel. |

## MALFUNCTIONS AND CORRECTIONS

| Cause | Correction |
|---|---|

**d. Automatic Loader Feed Rollers Fail To Rotate.**

| | |
|---|---|
| Feed roller catch head taper pin out of position. | Notify ordnance maintenance personnel. |

**e. Gun Breech Top Cover Damaged or Broken.**

| | |
|---|---|
| Failure to latch top cover before firing. | Notify ordnance maintenance personnel. |
| Breech ring barrel catch not engaging slot in tube, thus preventing the proper closing and latching of top cover. | Notify ordnance maintenance personnel. |

## 53. RECOIL MECHANISM.

**a. Recoil Violent or Excessive in Length.**

| | |
|---|---|
| Insufficient liquid in recoil cylinder. | Refill recoil cylinder as prescribed (par. 66 c). |
| Weak or broken recuperator spring. | Change barrel assemblies. If this corrects condition, refer original barrel assembly to ordnance maintenance personnel for replacement of recuperator spring. |

**b. Recoil Slow.**

| | |
|---|---|
| Breech ring guides and guideways in breech casing lack lubrication or are dirty. | Lubricate properly. Remove dirt and congealed oil. |
| Burs on breech ring guides and guideways of breech casing. | Notify ordnance maintenance personnel. |
| Barrel guide sleeve and locking collar loose. | Tighten barrel guide sleeve locking collar. Tightly screw in barrel guide sleeve locking collar set screw and barrel locking collar set screw friction disk. |
| Barrel guide sleeve burred or damaged. | Notify ordnance maintenance personnel. |

**c. Counterrecoil Slow.**

| | |
|---|---|
| Breech ring guides and guideways in breech casing lack lubrication or are dirty. | Lubricate properly. Remove dirt and congealed oil. |
| Damaged or burred breech ring guides and guideways of breech casing. | Notify ordnance maintenance personnel. |
| Incorrect setting of control rod valve spindle. | Adjust setting of spindle as required (par. 38 b). |

TM 9-252
53-54

## 40-MM AUTOMATIC GUN M1 (AA) AND 40-MM ANTIAIRCRAFT GUN CARRIAGES M2 AND M2A1

| Cause | Correction |
|---|---|
| Weak or broken recuperator spring. | Change barrel assemblies. Refer original barrel assembly to ordnance maintenance personnel. |
| Feed rods binding in automatic loader. | Notify ordnance maintenance personnel. |

d. **Counterrecoil Violent.**

| | |
|---|---|
| Loss of liquid in recoil cylinder. | Refill recoil cylinder as prescribed (par. 66 c). |
| Incorrect setting of control rod valve spindle. | Adjust setting of spindle as required (par. 38 b). |

e. **Failure to Counterrecoil.**

| | |
|---|---|
| Control rod valve spindle closed. | Adjust setting of spindle as required (par. 38 b). |
| Damaged or burred breech ring guides or guideways in breech casing. | Notify ordnance maintenance personnel. |
| Weak or broken recuperator spring. | Change barrel assemblies. Refer original barrel assembly to ordnance maintenance personnel. |
| Loading tray jammed on automatic loader feed rods. | Notify ordnance maintenance personnel. |

## 54. ELEVATING AND TRAVERSING MECHANISMS.

a. **Backlash in Elevating and Traversing Mechanisms.**

| | |
|---|---|
| Wear and improper meshing of worm, wormwheel, gears, or traversing arc; improper assembly of gears; end play in gear shafts. | Notify ordnance maintenance personnel. |

b. **Excessive Handwheel Effort Necessary To Elevate and Depress.**

| | |
|---|---|
| Equilibrators out of adjustment. | Adjust equilibrators (par. 67 b). |
| Equilibrator spring or springs weak or broken. | Notify ordnance maintenance personnel. |
| Parts improperly mated, worn, or damaged; lack of lubricant; dirty or gummy lubricant; elevating arc binding; ball bearings "locking up." | Notify ordnance maintenance personnel. |

126

## MALFUNCTIONS AND CORRECTIONS

### 55. FRONT AND REAR AXLES.

**a. Front Axle Loose in Swivel; Rear Axle Loose in Bearings.**

| Cause | Correction |
|---|---|
| Chassis compensating spring lock and arc disengaged, worn, or damaged. Lock broken due to excessive pressure on handle when not properly engaged. Ice in housing preventing operation of lock. | Keep locking handle in locked position and secured with strap and clip; wire handle down if necessary. Notify ordnance maintenance personnel (par. 67 e). |
| Loose or worn bearings in swivel or rear girder. | Notify ordnance maintenance personnel. |

**b. Axles, Cross Heads, Springs, Shackles, or Spindles Bent or Broken.**

| Cause | Correction |
|---|---|
| Heavy road shocks. | Notify ordnance maintenance personnel. |

### 56. WHEELS AND TIRES.

**a. Front Wheel Shimmy.**

| Cause | Correction |
|---|---|
| Looseness or excessive wear in steering mechanism. | Notify ordnance maintenance personnel. |
| Loose or broken wheel bearings. | Adjust bearings. If damaged or broken, notify ordnance maintenance personnel. |
| Dragging front wheel brakes. | Notify ordnance maintenance personnel. |

**b. Excessive Tire Wear.**

| Cause | Correction |
|---|---|
| Misalinement of wheels. | Notify ordnance maintenance personnel. |
| Wheels bent or loose on hubs; wheel bearings loose. | Adjust bearings. Tighten loose wheel retaining nuts. Replace bent wheels. |
| Low inflation pressure. | Keep tires inflated to recommended pressure, 45 pounds per square inch. |

### 57. ELECTRIC BRAKES.

**a. No Braking, or Intermittent Braking.**

| Cause | Correction |
|---|---|
| Broken wire in circuit. | Check entire wiring for broken wires. Repair or replace. |
| Broken wire in magnet. | If broken wire is on outside of magnet, repair if possible. If no current flows through magnet, notify ordnance maintenance personnel. |

TM 9-252
57

## 40-MM AUTOMATIC GUN M1 (AA) AND 40-MM ANTIAIRCRAFT GUN CARRIAGES M2 AND M2A1

| Cause | Correction |
|---|---|
| Controller defective. | Notify ordnance maintenance personnel. |
| Poor connections. | Check, clean, and tighten all connections at brake, controller, load control, and sockets. |
| Poor ground condition in circuit. | Clean up and tighten all connections. |
| Defective plug or socket. | Check for loose connections, dirty or corroded blades, or broken socket. Repair, or replace with new socket. |

**b. Very Weak Brakes.**

| | |
|---|---|
| Worn out or greasy brake lining. | Lining may be worn to full extent of magnet travel. Refer to ordnance maintenance personnel for new lining. |
| Glazed magnet facing. | Roughen facing of magnet with PAPER, flint, class B, No. 1. |
| Foot control out of adjustment on prime mover. | When carriage brakes are adjusted, effective range of controller is changed. This throws controller out of adjustment. Reset to cover full range of controller. |
| Wire broken in insulation; bare wire; loose connection; poor contact at load control. | Check wiring and connections for defects. Short out load control. Correct if possible; otherwise, notify ordnance maintenance personnel. |
| Insufficient current. | Test with ammeter (par. 67 i (2)). Clean up and tighten all connections. Check plug and socket for corroded or dirty blades, or broken socket. Replace broken parts. |
| Stop lights connected in circuit. | Check to determine if stop lights have been connected in circuit by mistake. |
| Poor ground connections. | Ground contacts must be solid and clean. |
| Worn wheel bearings. | Refer to ordnance maintenance personnel. |

**TM 9-252**
**57**

## MALFUNCTIONS AND CORRECTIONS

c. Brakes Grabbing.

| Cause | Correction |
|---|---|
| Loose or worn wheel bearings. | Tighten bearings. If this does not correct condition, refer to ordnance maintenance personnel for replacement of bearings. |
| All brakes not working. | Check current at brakes with ammeter (par. 67 l (2)). Check for broken wires and poor connections. |
| Sticky or grease-coated lining. | Wash lining with SOLVENT, dry-cleaning. |
| Contactor arm of controller pitted. | Smooth out contactor arm with PAPER, flint, class B, No. 1. |
| Stop lights in brake circuit. | Determine if stop lights have been connected improperly. |
| Poor electrical connections. | Check wiring for loose connections, bare wires, or wires broken in insulation. |
| Worn lining; drums out of round; lining loose on rivets; broken or weak band or magnet springs; controller burned out; poor contactor blade spacing; magnet bushing worn. | Refer to ordnance maintenance personnel. |

d. Brakes Dragging.

| | |
|---|---|
| Drums out of round; broken spring in hand control; insufficient spacing between armature and magnet; band distorted; unequal clearance; insufficient lining clearance; loose or damaged wheel bearings. | Refer to ordnance maintenance personnel. |

e. Battery on Carriage Runs Down.

| | |
|---|---|
| Safety switch left in "ON" position. | Keep safety switch lever in "OFF" position. |
| Excessive use of battery to apply brakes. | Do not use battery current for parking; use hand brakes. Use battery current for braking only as necessary. |
| Short circuit in brake system. | Check for bare wire. Tape or replace damaged cable. |

**40-MM AUTOMATIC GUN M1 (AA) AND 40-MM ANTIAIRCRAFT GUN CARRIAGES M2 AND M2A1**

## 58. LIGHTS.

a. **Dim Lights.**

| Cause | Correction |
|---|---|
| Loose or corroded connections. | Clean and tighten all connections. |
| High resistance in blackout light switch. | Notify ordnance maintenance personnel. |
| Damaged wires. | Check for broken or bare wires. |

b. **Flickering Lights.**

| Cause | Correction |
|---|---|
| Loose or damaged plug or socket. | Check plug and socket. Tighten. Replace if broken. |
| Loose or bare wires; loose connections; loose bulbs. | Check wiring and connections. Cover or replace bare wires. Tighten loose connections and bulbs. |

---

Section VI

# LUBRICATION

## 59. INTRODUCTION.

a. **General.** Lubrication is an essential part of preventive maintenance, determining to a great extent the serviceability of parts and assemblies. Satisfactory operation and long life of the materiel are not assured unless the materiel is kept clean and well lubricated.

b. Apply sufficient lubricants, but avoid wasteful practices. Excessive lubrication will result in dust accumulations on some moving parts and, if not removed, may cause wear and malfunctioning. Particular attention should be given to the lubrication of sliding surfaces which contain no oilholes, plugs, or fittings. Keep all exposed parts clean and well lubricated. The materiel should always be lubricated after washing.

c. **Supplies.** In the field, it may not be possible to supply a complete assortment of lubricants called for by the lubrication guide to meet the recommendations. It will be necessary to make the best use of those available, subject to inspection by the officer concerned, in consultation with responsible ordnance personnel.

## 60. LUBRICATION GUIDE.

a. War Department Lubrication Guide No. 61 (figs. 108 and 109) prescribes first and second echelon lubrication maintenance. Lubrication to be performed by ordnance maintenance personnel is covered in TM 9-1252 and TM 9-1253.

## LUBRICATION

*h.* A lubrication guide is placed on or is issued with each item of materiel, and is to be carried with it at all times. In the event the materiel is received without a guide, the using arms shall immediately requisition a replacement from the Commanding Officer, Fort Wayne Ordnance Depot, Detroit 32, Michigan.

*c.* Lubrication instructions on the guide are binding on all echelons of maintenance and there shall be no deviations, except as indicated in subparagraph d, below.

*d.* Service intervals specified on the guide are for normal operating conditions. These intervals will be reduced under extreme conditions such as excessively high or low temperatures, prolonged periods of operation, continued operation in sand or dust, immersion in water or exposure to moisture, any one of which may quickly destroy the protective qualities of the lubricant.

*e.* Lubricants are prescribed in the "Key" in accordance with three temperature ranges, "above $+32$ F," "$+32$ F to 0 F," and "below 0 F." When to change grades of lubricants is determined by maintaining a close check on operation of the materiel during the approach to change-over periods, especially during initial action. Sluggish action is an indication of lubricants thickening and the signal to change to grades prescribed for the next lower temperature range. Ordinarily it will be necessary to change grades of lubricants *only when air temperatures are consistently in the next higher or lower range,* unless malfunctioning occurs sooner due to lubricants being too thin or too heavy.

*f. Lubrication Equipment.*

(1) Each piece of materiel is supplied with lubrication equipment adequate to maintain the materiel. This equipment will be cleaned both before and after use.

(2) Lubrication guns will be operated carefully and in such manner as to insure a proper distribution of the lubricant.

*g. Points of Application.*

(1) Lubrication fittings, grease cups, oilers, and oilholes are readily identifiable on the materiel by a red circle. Such lubricators and the surrounding surface will be wiped clean before lubricant is applied.

(2) Where relief valves are provided, apply new lubricant until the old lubricant is forced from the vent. Exceptions are specified in notes on the lubrication guide.

*h. Cleaning.* SOLVENT, dry-cleaning, or OIL, fuel, Diesel, will be used to clean or wash all parts. Use of gasoline for this purpose is prohibited. After washing, parts will be thoroughly dried before applying lubricant. Swab gun bore with solution of ½ pound of SODA ASH to each gallon of warm water, or with a thick suds of issue soap and warm water. Rinse with clear water and dry thoroughly before oiling.

**TM 9-252**
**60**

### 40-MM AUTOMATIC GUN M1 (AA) AND 40-MM ANTIAIRCRAFT GUN CARRIAGES M2 AND M2A1

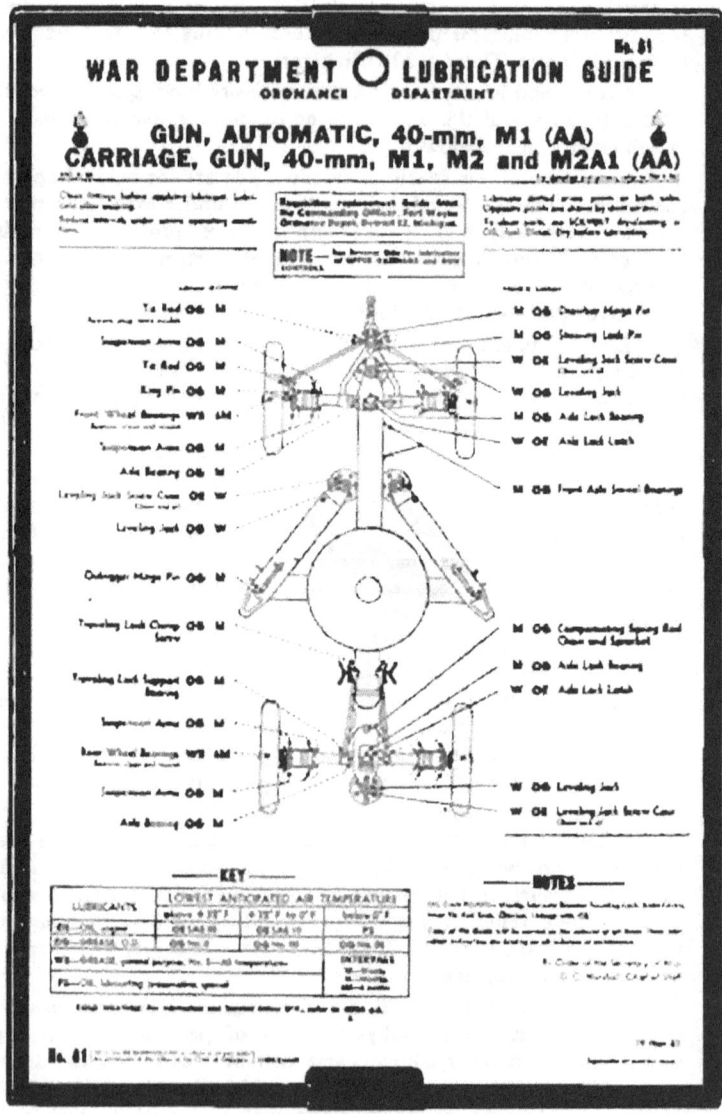

*Figure 108 — Lubrication Guide*

LUBRICATION

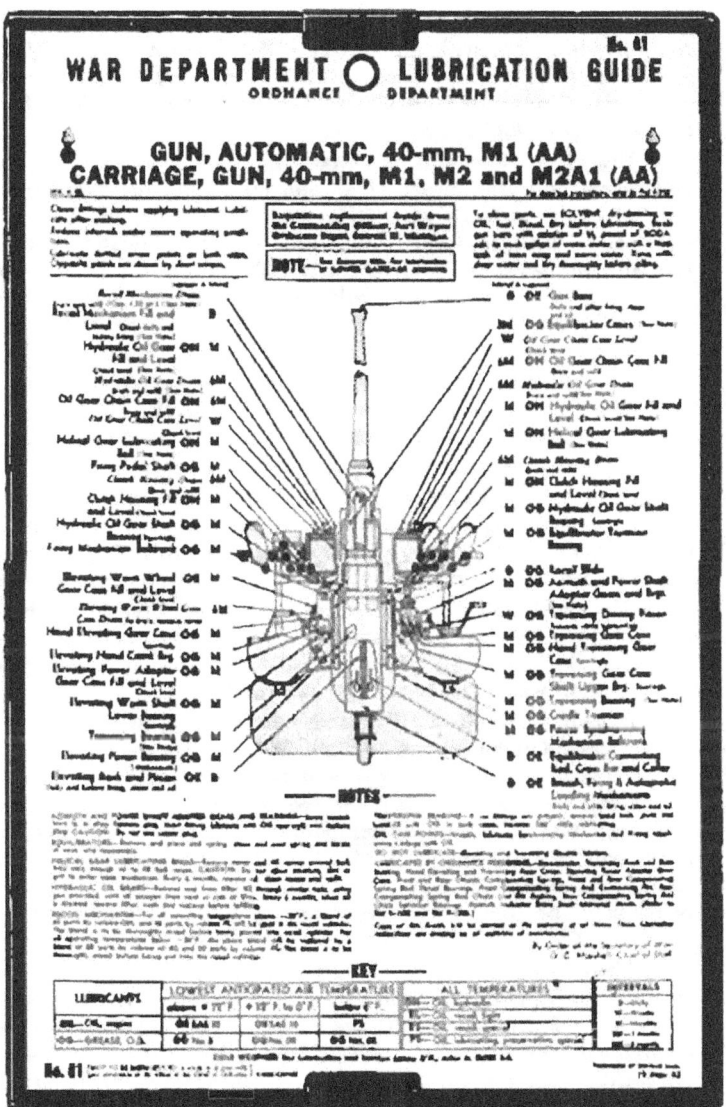

Figure 109 — Lubrication Guide

**TM 9-252**

**40-MM AUTOMATIC GUN M1 (AA) AND 40-MM ANTIAIRCRAFT GUN CARRIAGES M2 AND M2A1**

i. Azimuth and Power Shaft Adapter Gears and Bearings. Monthly, lubricate with GREASE, O.D., No. 0, above +32 F or GREASE, O.D., No. 00, below +32 F. Some models have ¼-inch plug. To lubricate, remove plug and insert fitting. CAUTION: Do not use upper plug or overlubricate.

j. Breech, Firing, and Automatic Loading Mechanisms. Daily and before and after firing, clean and oil all moving parts and unpainted metal surfaces with OIL, engine, SAE 30, above +32 F; OIL, engine, SAE 10, from +32 F to 0 F; OIL, lubricating, preservative, special, below 0 F. CAUTION: To insure easy breech operation and to avoid misfiring in cold weather, clean, dry, and lubricate with OIL, lubricating, preservative, special. To clean firing mechanism, remove and operate pin in SOLVENT, dry-cleaning.

k. Equilibrators. Every 3 months, remove end piece and spring; clean and coat spring and inside of case with GREASE, O.D., No. 0, above +32 F, or GREASE, O.D., No. 00, below +32 F. CAUTION: Use equilibrator spring compressor to remove spring.

l. Helical Gear Lubricating Balls. Remove cover and fill recess around ball with OIL, hydraulic. Add only enough oil to fill ball recess. CAUTION: Do not allow moisture, dirt, or grit to enter open mechanism. Every 6 months, remove oil, clean recess, and refill.

m. Hydraulic Oil Gears. Remove cap from filter, fill with OIL, hydraulic, through center hole, using gun provided, until oil escapes from vent at side of filter. Every 6 months, when oil is drained, remove filter, wash, and replace before refilling.

n. Recoil Mechanism. For all operating temperatures above −20 F, a blend of 60 parts by volume of OIL, hydraulic, and 40 parts by volume of OIL, recoil, light, will be used in the recoil cylinder. The blend is to be thoroughly mixed before being placed in recoil cylinder. For all operating temperatures below −20 F, the above blend will be replaced by a blend of 50 parts by volume of OIL, recoil, special, and 50 parts by volume of OIL, recoil, light. The blend is to be thoroughly mixed before being put into the recoil cylinder. Capacity 1.32 pints.

o. Traversing Bearing. Monthly, lubricate traversing bearing with GREASE, O.D., No. 0, above +32 F or GREASE, O.D., No. 00, below +32 F through the fittings. If no fittings are present, remove handhole plate and hand-fill with GREASE, O.D., No. 0, above +32 F, GREASE, O.D., No. 00, below +32 F. In both cases, traverse 360 degrees while lubricating.

p. Wheel Bearings. Remove bearing cone assemblies from hub. Wash bearings, cones, spindle, and inside of hub, and dry thoroughly. Do not use compressed air. Inspect bearing races and replace if damaged. Wet the spindle and inside of hub and hub cap with

**LUBRICATION**

GREASE, general purpose, No. 2, to a maximum thickness of $\frac{1}{16}$ inch only to retard rust. Lubricate bearings with GREASE, general purpose, No. 2, with a packer, or by hand, kneading lubricant into all spaces in the bearing. Use extreme care to protect the bearings from dirt, and immediately reassemble and replace wheel. Do not fill hub or hub cap. The lubricant in the bearing is sufficient to provide lubrication until the next service period. Any excess might result in leakage into the drum. Adjust bearings in accordance with instructions in paragraph 88 d.

q. **Oil Can Points.** Weekly, lubricate power synchronizing mechanism and firing mechanism linkage, drawbar traveling lock, brake levers, inner tie rod ends, clevises, linkage with OIL, engine, SAE 30, above +32 F, or OIL, engine, SAE 10, from +32 F to 0 F; use OIL, lubricating, preservative, special, below 0 F.

r. **Do Not Lubricate.** Elevating and traversing electric motors.

s. **Lubricated by Ordnance Personnel.**

(1) TRAVERSING RACK AND BASE BEARING. Every 6 months, disassemble, clean, and lubricate with GREASE, O.D., No. 0, above +32 F or GREASE, O.D., No. 00, below +32 F.

(2) HAND ELEVATING AND TRAVERSING GEAR CASES AND ELEVATING POWER ADAPTER GEAR CASE. Every 6 months, disassemble; clean housing and enclosed parts. Repack with GREASE, O.D., No. 0, above +32 F or GREASE, O.D., No. 00, below +32 F.

(3) RECUPERATOR. Every 6 months, remove recuperator, disassemble, and apply a thin coat of GREASE, O.D., No. 0, above +32 F or GREASE, O.D., No. 00, below +32 F to spring and inside of case.

(4) FRONT CHASSIS COMPENSATING SPRING, FRONT COMPENSATING SPRING ROD THRUST BEARING, AND FRONT COMPENSATING SPRING ROD CONNECTING PIN. Every 6 months, remove compensating spring and spring rod thrust bearing from compensating spring housing and clean all parts. Repack bearing and coat spring. Coat inside of spring housing and swivel body. Use GREASE, O.D., No. 0, above +32 F or GREASE, O.D., No. 00, below +32 F. Remove axle lock and cover assembly to facilitate removal of plug from compensating spring rod connecting pin. Insert fitting and lubricate with GREASE, O.D., No. 0, above +32 F, or GREASE, O.D., No. 00, below +32 F. Before replacing cover, wipe the compensating locating arc clean and recoat with GREASE, O.D., No. 0, above +32 F or GREASE, O.D., No. 00, below +32 F.

(5) REAR CHASSIS COMPENSATING SPRING, REAR COMPENSATING SPRING ROD THRUST BEARING, REAR COMPENATING SPRING ROD CHAIN LINK PIN BUSHING, AND REAR COMPENSATING SPRING ROD CHAIN SPROCKET BEARINGS. Every 6 months, remove rear compensating spring unit and chain from chassis frame. Disassemble and clean all parts. Repack spring rod thrust bearing and coat compensating

TM 9-252
60-62

**40-MM AUTOMATIC GUN M1 (AA) AND 40-MM ANTIAIRCRAFT GUN CARRIAGES M2 AND M2A1**

spring and inside of spring housing with GREASE, O.D., No. 0, above +32 F or GREASE, O.D., No. 00, below +32 F. Apply 6 or 8 drops of OIL, engine, SAE 30, above +32 F, OIL, engine, SAE 10, from +32 F to 0 F; or OIL, lubricating, preservative, special, below 0 F to each roller of the compensating spring rod chain. Work the oil in between the rollers and connecting links and wipe off excess. Disassemble the chain sprocket bearings, clean, and repack with GREASE, O.D., No. 0, above +32 F or GREASE, O.D., No. 00, below +32 F. Wipe the compensating arc clean, coat teeth, and lubricate the compensating rod chain link pin bushing through fitting with GREASE, O.D., No. 0, above +32 F or GREASE, O.D., No. 00, below +32 F. Replace covers and lubricate the compensating spring rod chain and sprocket as prescribed in the lubrication guide.

(6) AZIMUTH INDICATOR DRIVE SHAFT UNIVERSAL JOINTS. Every 6 months, remove, clean, and repack with GREASE, O.D., No. 0, above +32 F or GREASE, O.D., No. 00, below +32 F.

## 61. REPORTS AND RECORDS.

a. Reports. Unsatisfactory performance of materiel will be reported to the ordnance officer responsible for maintenance.

b. Records. A complete record of seasonal changes of lubricants and recoil oils will be kept in the artillery gun book for the materiel.

---

### Section VII

## CARE AND PRESERVATION

### 62. GENERAL.

a. It is of vital importance that all parts of the materiel be kept in proper condition in order that the weapon be ready for immediate service at all times. Lubricating, cleaning, and preserving materials provided with the gun and carriage will enable the personnel to keep the parts in proper working condition. This section of the manual prescribes the uses of these materials. Section VI prescribes lubrication instructions for normal conditions. Section XIV prescribes special precautions to be taken for operation under unusual conditions.

b. Moving parts of the various mechanisms should be lubricated in the prescribed manner, and periodical examinations should be made to insure that the lubricant is reaching the parts for which it is intended, and to insure that the weapon as a whole is receiving the attention necessary for satisfactory operation. Vigilance should be exercised by all members of the using arm personnel to note and report any irregularity in the operation or functioning of the gun and mount.

## CARE AND PRESERVATION

c. Dirt and grit, accumulated in traveling or from the blast of the piece in firing, settle on bearing surfaces, and in combination with the lubricant itself, form a cutting compound. Powder fouling attracts moisture and hastens the formation of rust. It is essential that all parts be cleaned at frequent intervals, depending upon use and service.

d. If rust should accumulate, its removal from bearing surfaces requires special care in order that clearances shall not be unduly increased. CLOTH, crocus, should be used for this purpose. The use of coarse abrasives is strictly forbidden.

e. In disassembly, assembly, or inspection, extreme care must be exercised to prevent dust, dirt, or other foreign matter from entering the mechanisms.

f. When materiel is not in use, the proper covers must be used.

g. When the weapon is to be unused for more than one week, the bore, breech mechanism, and bright unpainted surfaces should be cleaned with SOLVENT, dry-cleaning, and the surfaces coated with COMPOUND, rust-preventive, light. Parts coated with this compound should be examined weekly and the compound renewed if necessary.

h. Should an enemy shell or bomb burst near the weapon, it must be determined that the weapon has not been damaged to a dangerous degree before the next round is fired. Damage of a serious nature should be reported to the ordnance officer.

## 63. ORGANIZATION SPARE PARTS AND ACCESSORIES.

a. All organization spare parts, tools, and accessories should be kept in an orderly manner so that they can be located quickly when required. They should be protected from loss or damage by being kept in their proper rolls and chests. Those items susceptible to rust and corrosion must be cleaned thoroughly at regular intervals and coated with a film of oil. Parts supplied in protective containers should be kept in the containers until required.

b. The sets of organization spare parts and accessories for the gun and carriage should be maintained as completely as possible at all times. The sets should be checked with the lists in the Standard Nomenclature Lists (sec. XV), and all used parts, and all missing parts, tools, and accessories should be replaced immediately.

## 64. CLEANING AFTER FIRING.

a. Bore. The bore of the barrel should be thoroughly cleaned, rinsed, dried, and oiled immediately after firing and while the gun is still warm. This procedure is to be repeated each day until sweating ceases. The purpose of the repeated cleanings is to remove the effects of sweating, a chemical reaction of the burned powder composition which cannot be removed with the initial procedure.

TM 9-252
64

### 40-MM AUTOMATIC GUN M1 (AA) AND 40-MM ANTIAIRCRAFT GUN CARRIAGES M2 AND M2A1

(1) The cleaning procedure can be done most easily if the barrel is removed. In this case, the cleaning operations are conducted from the breech end of the barrel. The barrel can be cleaned while mounted in place with the breechblock and extractors removed. Such cleaning operations must be conducted from the muzzle end of the barrel.

(2) CLEAN.

(a) For cleaning, use a solution made by dissolving ½ to 1 pound of SODA ASH in 1 gallon of boiling water, or CLEANER, rifle bore, or a solution of boiling water and issue soap.

CAUTION: Do not use SOLVENT, dry-cleaning.

(b) Assemble the three sections of the cleaning staff M14. Wrap the end of the staff with pieces of cloth soaked in the solution or cleaner. With two men on the staff, work the staff through the bore, using a pushing and pulling action. A pad or swab may be made of cloths, soaked in the solution, and pushed repeatedly through the bore. Wash the bore thoroughly in this manner.

NOTE: In the absence of SODA ASH, CLEANER, rifle bore, or issue soap, hot water may be used alone. SOLVENT, dry-cleaning, should not be used because the corrosive salts from the powder composition, which cause rust, are not readily dissolved by petroleum products; they are readily dissolved by water solutions.

(3) RINSE. Follow the cleaning operation by rinsing the bore to remove the cleaning solution. Wrap the end of the staff with cloths soaked in clean, hot water, or use swabs soaked in clean, hot water. Follow the procedure described for cleaning.

(4) DRY. Follow the procedure prescribed for cleaning, using the same accessories and methods but substituting dry cloths for wet ones.

(5) INSPECT. Inspect the barrel for flaws, cracks, or other defects which would make it unfit for further service. Report any defects to ordnance maintenance personnel.

(6) OIL. Assemble the bore brush M29 to the cleaning staff. Place the brush over a bucket to catch the oil which drips from the brush and apply oil to the brush. Use OIL, engine, SAE 30, for temperatures above 32 F, or OIL, engine, SAE 10, for temperatures below 32 F. Apply a coating of oil to the bore.

NOTE: Special instructions for cleaning the gun at temperatures below 32 F are given in paragraph 123.

b. Barrel Exterior. With a clean cloth, apply a coating of light oil to all exterior parts not painted or otherwise protected against rusting.

c. Breech Mechanism. After firing, the breechblock, inner cranks, and extractors should be removed (par. 80), the breech-

## CARE AND PRESERVATION

block should be disassembled (par. 82), and all parts should be cleaned and oiled thoroughly, assembled, and installed. Special attention should be given to the firing pin and inner and outer cocking levers. The hole in the firing pin bushing or in the front of the breechblock must be checked to insure that it is clean and unobstructed. The protrusion of the firing pin through this hole in firing position must be checked frequently with the striker protrusion gage. Any malfunction, weak or broken spring, or broken or deformed part should be corrected or replaced, if this work can be performed by the using arm personnel; otherwise, the condition should be reported to the ordnance maintenance personnel.

65. GUN.

a. Barrel Assembly.

(1) Guns become copper-fouled to less extent when cared for in the proper manner. Wear in the bore does not depend entirely upon the number of rounds fired but also on the care given the bore in cleaning and cooling between periods of firing.

(2) Since the accuracy life of a gun is decreased by a fast rate of firing and the attendant heat, the gun should be allowed to cool and should be washed as often as practical. Barrel assemblies should be changed during firing whenever necessary because of overheating. Each projectile must be cleaned before it is placed in the rack preparatory to firing. The bore should be cleaned before firing.

(3) It is important that, whenever possible, the gunner inspect the bore to make certain that it does not contain any extraneous particles that might cause damage to the gun.

(4) Make certain that the barrel guide sleeve and locking collar are properly positioned at all times, and that the barrel guide sleeve locking collar set screw and the barrel locking collar set screw friction disk are tightly screwed into place. If this set screw is not properly tightened, the barrel guide sleeve and locking collar have a tendency to work loose. This condition will allow the recuperator spring to expand and will result in faulty recoil and counterrecoil.

(5) It will be the responsibility of battery personnel to see that the barrel guide sleeve and locking collar are properly positioned at all times, and that the barrel guide sleeve locking collar set screw, and the barrel locking collar set screw friction disk, are tightly screwed into place.

(6) The recuperator spring collar wrench B200473 is to be used in tightening the barrel guide sleeve locking collar, and may be obtained by requisition through regular channels on a basis of one per battery.

(7) The flash hider cover should be in place when the gun is not being fired.

CAUTION: Copper fouling must not be removed under any circumstances.

## 40-MM AUTOMATIC GUN M1 (AA) AND 40-MM ANTIAIRCRAFT GUN CARRIAGES M2 AND M2A1

b. **Regular Cleaning Procedure.**

(1) When the gun is not being fired, this cleaning procedure is to be followed at intervals specified by the officer in charge. The interval will be dependent upon atmospheric, traveling, or other conditions.

(2) This cleaning procedure is the same as that specified in paragraph 64 a, except that SOLVENT, dry-cleaning (which readily removes oil and dirt), is used in place of the SODA ASH or issue soap solutions.

c. **Inactivity.** If the weapon is to be unused for more than a week, the bore should be given a coating of COMPOUND, rust-preventive, light.

d. **Breech Mechanism.**

(1) At frequent intervals, when the gun is not being fired, the breech mechanism will be disassembled, cleaned, oiled, and assembled. These intervals will be specified by the officer in charge, and their frequencies will be dependent upon atmospheric and other conditions.

(2) The automatic loader hood and shield should be in place when the gun is not actually in use, to prevent dust and grit from getting into the mechanism, causing wear, and impeding smooth operation. Covers must be kept closed and locked.

(3) If the breech mechanism does not operate smoothly, or if the mechanism requires a greater effort than usual when operated with the hand operating lever, it should be disassembled and the cause determined. If corrective measures are beyond the scope of the using arm personnel, the matter should be brought to the attention of the ordnance maintenance personnel. Any part showing signs of scoring or burs must be replaced.

e. **Barrel Assembly Removal.** The barrel assembly may be removed from the breech ring by using arm personnel (par. 79). The barrels of the 40-mm Automatic Gun M1 (AA) are interchangeable.

## 66. RECOIL MECHANISM.

a. **General.**

(1) It is extremely important that the recoil cylinder be filled in the prescribed manner and that the recoil mechanism be correctly prepared for action. Should these operations be performed incorrectly, should the recoil cylinder anchor bracket be mounted in reverse position (par. 86 f), or should there be an appreciable loss of liquid, unsatisfactory recoil or counterrecoil must be expected and serious damage to the mechanism may result. To insure that the recoil mechanism is kept in good order, it is important that it be cared for by trained personnel.

(2) The energy of recoil is absorbed mainly in the recoil cylinder although a certain amount of the energy is taken up and stored by the

## CARE AND PRESERVATION

recuperator spring on the tube. Additional energy is absorbed in overcoming the friction in the slides and packing.

(3) The recoil energy taken up and stored by the recuperator spring is used to return the gun into battery in counterrecoil. The speed of counterrecoil is controlled by the counterrecoil buffing action of the recoil cylinder. This is governed by the counterrecoil adjusting valve and adjusted by turning the squared head of the control rod valve spindle which protrudes from the front end of the recoil cylinder.

(4) The *length* of recoil cannot be adjusted by turning the control rod valve spindle; this merely adjusts speed of counterrecoil.

b. Length of Recoil.

(1) The recoil should be smooth and the length of recoil should be between 7.4 and 8.3 inches. The most desirable length is 7.83 inches. The method of measuring the length of recoil is described in paragraph 44 b. The length of recoil is measured while the gun is being fired by means of the recoil indicator.

(2) The extreme length of recoil is fixed by the dimension between the recoil cylinder piston and the rear end of the recoil cylinder which forms a seating for the packing. Maximum recoil is approximately 8.8 inches. This length, however, should not be reached, as a recoil of such length will result in damage to the mechanism.

(3) If the length of recoil does not come within the proper limits, check the fluid level in the recoil cylinder. Fill to the proper level, if necessary. If recoil is unsatisfactory with a properly filled recoil cylinder, the cause may be a weak or broken recuperator spring; change barrel assemblies.

(4) If the length of recoil is corrected by changing barrels, this would indicate that the recuperator spring on the original barrel was unsatisfactory. This must be reported to ordnance maintenance personnel for checking and replacement if found necessary.

(5) If changing barrel assemblies does not correct the length of recoil, the weapon should be reported to ordnance maintenance personnel.

c. Recoil Cylinder Filling.

(1) The recoil cylinder is filled with a mixture of 60 parts by volume of OIL, hydraulic, and 40 parts by volume of OIL, recoil, light. Below 20 F, a mixture of 50 percent OIL, recoil, light, and 50 percent OIL, recoil, special, will be used. The capacity of the recoil cylinder is 1.32 pints. The blend will be mixed thoroughly in a clean, dry container. The method of filling provides for leaving a small air space when the recoil cylinder is correctly filled.

(2) Under no circumstances should this mixture be added to the glycerine-water mixture specified for use previously. Before using the new mixture to replace the glycerine-water mixture, drain the recoil cylinder, disassemble the mechanism, and clean thoroughly.

TM 9-252
66

**40-MM AUTOMATIC GUN M1 (AA) AND 40-MM ANTIAIRCRAFT GUN CARRIAGES M2 AND M2A1**

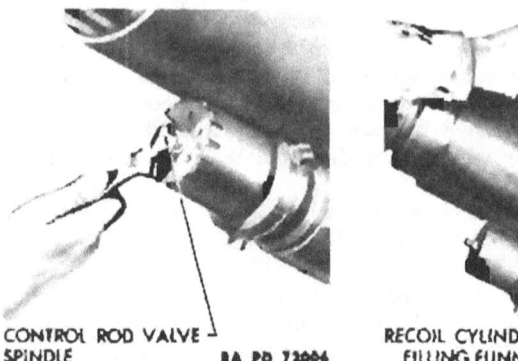

Figure 110 — Adjusting Control Rod Valve Spindle

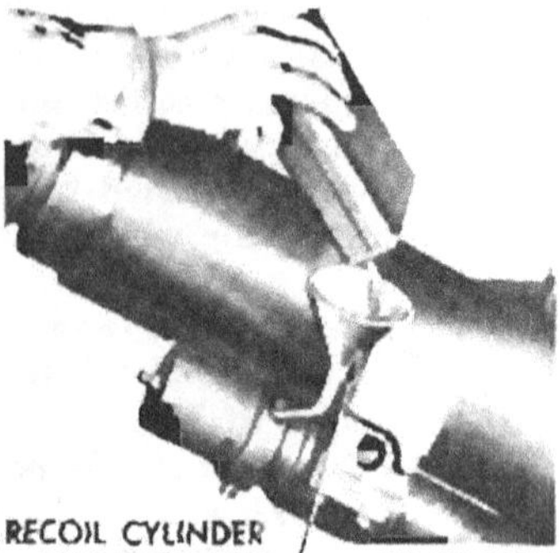

Figure 111 — Replenishing Fluid in Recoil Cylinder

The change-over to the new mixture will be performed only by ordnance maintenance personnel.

(3) To fill the recoil cylinder, proceed as follows:

(a) Elevate the gun to 25 degrees. Remove the recoil cylinder filling plug. Remove the control rod valve spindle adjustment screw and plate. Open the control rod valve by turning the squared end of the control rod valve spindle (fig. 110). Using the filling funnel, pour the blend of recoil and hydraulic oil in until it overflows from the filling valve (fig. 111). Remove the funnel and replace the filling plug.

(b) Elevate and depress the gun slowly a few times, finishing with the gun at 25 degrees elevation. Remove the filling plug again and, using the funnel, pour in liquid until full. Reset the control rod valve by screwing the control rod valve spindle down completely and then unscrewing it one-third turn. Replace the adjusting plate and screw. Replace the filling plug.

d. **Recoil Cylinder Emptying.** Depress the gun. Open the control rod valve. Remove the filling plug. Place a clean, dry pail or any other receptacle in position and remove the draining plug.

e. **Recoil Cylinder Exercising.**

(1) The recoil cylinder piston rod will corrode and stick at the rear packing gland unless it is exercised frequently. Such sticking may result in severe damage to the weapon if the gun is fired. The recoil cylinder piston rod must be exercised weekly to insure its proper operating condition.

(2) Detach the recoil cylinder from the breech casing (par. 81 c). With the piston rod retaining pin held in the lugs on the front of

## CARE AND PRESERVATION

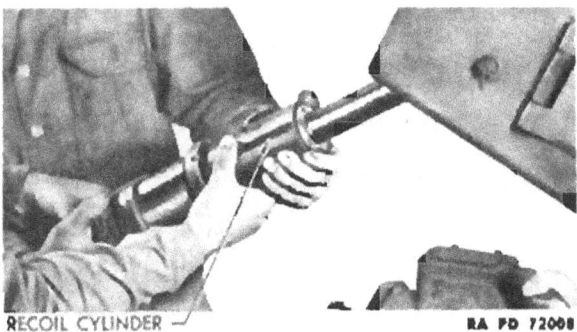

*Figure 112 — Exercising Recoil Cylinder*

the breech ring, pull the piston rod to its limit out of the recoil cylinder (fig. 112) and return it to its original position. Do this several times.

(3) Should the piston rod be corroded, it may be very difficult to break it free. In this case, assemble the recoil cylinder to the breech casing in the reversed position (recoil rod pointing toward gun muzzle). This will provide a means of holding the recoil cylinder without danger of crushing it. Tap or apply pressure to the retaining pin to turn the rod on its own axis to break it free.

(4) After the piston rod has been freed, polish the corroded area with CLOTH, crocus, and oil lightly. Install the recoil cylinder.

NOTE: If the battery personnel cannot free the piston rod by the methods suggested, the mechanism should be referred to ordnance maintenance personnel for overhaul.

f. Care of Recoil Oil.

(1) A blend of 60 parts by volume of OIL, hydraulic, and 40 parts by volume of OIL, recoil, light, is used in the recoil mechanism. Below 20 F, a mixture of 50 percent OIL, recoil, light, and 50 percent OIL, recoil, special, will be used. Care must be taken not to use any liquid other than that prescribed. Water and other foreign matter must not be introduced into recoil mechanisms that use recoil oil.

(2) Recoil oil, hydraulic oil, or the correct mixture of these two oils must not be put into any container not marked with the name of the oil, nor left in open containers, nor be subjected to excessive heat, nor mixed with any other type of oil than those prescribed. The transfer of these liquids to containers not properly marked may result in the wrong liquid getting into the recoil mechanism, or in the use of recoil oil or the prescribed mixture being used for lubricating purposes.

## 40-MM AUTOMATIC GUN M1 (AA) AND 40-MM ANTIAIRCRAFT GUN CARRIAGES M2 AND M2A1

(3) When putting the mixture into the system, it should be filtered through a piece of clean cloth as well as through the wire strainer of the filling funnel. Every precaution must be taken to prevent the introduction of water or grit into the mechanism, either in the mixture or through failure to clean thoroughly the connections and servicing equipment.

(4) Exposure of these liquids in an open can may result in the accumulation of moisture. Condensation in a container partly filled with oil or the mixture, or the pouring of any of these liquids from one container to another which has moisture on its inner walls, results in moisture being carried along into the recoil mechanism.

(5) If there is a possibility that the recoil oil, hydraulic oil, or the blend of these two oils may contain water, the suspected liquid should be tested by one of these methods:

(a) Fill a clean glass container of pint capacity with the liquid. Permit the liquid to settle. The water being heavier will sink to the bottom, if present. With the container slightly tilted, drops or bubbles will form in the lower portion. Invert the container and hold it to the light. Drops or bubbles of water, if present, may be seen slowly sinking in the liquid. If the liquid has a cloudy appearance, the cloudiness may be ascribed to particles of water.

(b) Another test for water is to heat the liquid to 212 F (boiling point of water) in a shallow pan. Water in the liquid will appear on the surface as minute bubbles. This test will disclose water not determinable by the settling test.

(c) Should either test show water, the liquid on hand should be turned in.

## 67. CARRIAGE.

a. General.

(1) The care and preservation of the carriage in service requires proper cleaning, strict observance of the lubrication program, tightening of loosened parts, and repair or replacement of broken parts. When traveling, it calls for proper attachment of the gun stay, secure locking of the drawbar, front and rear wheel axles, steering link securing clamp, outriggers, leveling jacks, and accessories in their positions on the carriage, the full protective use of metal and cloth covers, the proper adjustment of wheel bearings and brakes, and the correct inflation of tires.

(2) All bearing surfaces, revolving parts, springs, screw threads, gear teeth, and exterior parts must be kept as clean and free from dirt as possible. Special attention must be given exposed teeth and bearing surfaces. Cotter pins must be properly spread. Bolts, nuts, and screws must be tight and locked td prevent their coming loose.

(3) The carriage should receive a general inspection periodically.

TM 9-252
67

## CARE AND PRESERVATION

(4) In adjusting, tightening, disassembling, and assembling the best available tools should be used. Care must be exercised that dirt, dust, or other foreign matter is not permitted to enter the mechanism.

CAUTION: The front and rear chassis compensating spring locks should remain locked during the removal of tires and wheels, and should not be unlocked until wheels and tires are replaced and leveling jacks are raised so that wheels are in solid contact with ground.

(5) Care must be exercised to detect any cutting or abrasion of the teeth of the elevating arc or other parts. Any deformation of this nature should be reported to the ordnance maintenance personnel for correction. Rust must not be permitted to accumulate on any part.

(6) When the carriage is to be stored, or is not to be used for more than a week, unpainted surfaces should be cleaned with SOLVENT, dry-cleaning, and coated with COMPOUND, rust-preventive, light.

(7) All lubricating fittings should be kept clean and, if necessary, a piece of wire may be used in cleaning out oil passages. Do not use a sliver of wood as it may splinter and clog the passage. Exercise care so as not to damage the interior of the fitting. In painting oil fittings, keep paint out of the openings.

(8) Remove drain plugs (there are two) and drain the chassis of the carriage as often as necessary to keep water from accumulating in the girder and the swivel body.

b. **Equilibrator Adjustment.**

(1) For proper operation of the remote control system, the equilibrators must be adjusted so that the effort required to elevate is the same as that required to depress the gun. Each equilibrator must be adjusted to the same degree to keep the pull from each equalized."

(2) Test the amount of effort required to elevate and depress the gun. If the effort to depress the gun is more than that required to elevate, adjust the equilibrators in this manner:

*(a)* Open both equilibrator covers. With the equilibrator rod bushing nut wrench, remove the equilibrator rod jam nuts from both rods. If the effort to depress the gun is more than that required to elevate, turn out the equilibrator rod bushing nuts; if the effort to elevate is the greater, turn in the bushing nuts. In tightening and loosening the nuts, be sure to adjust both exactly alike.

*(b)* When the adjustment is completed, lock both bushing nuts in position with their jam nuts. Close the end covers.

(3) The equilibrator cross bar and collar should be oiled frequently with OIL, engine, SAE 10, to prevent rusting. The trunnion bearings should be checked to make sure they are working smoothly and efficiently. The insides of the cases, springs, spacers, and other

## 40-MM AUTOMATIC GUN M1 (AA) AND 40-MM ANTIAIRCRAFT GUN CARRIAGES M2 AND M2A1

parts should be greased; see paragraph 87 for instructions for disassembly and assembly.

c. Elevating Rack and Pinion.

(1) The teeth of the elevating rack and pinion require little lubrication, but as a protection against rust, they must be covered with a thin coating of oil. Under normal conditions, dust and grit will adhere to this oily film and cause wear of both rack and pinion; consequently, the teeth must be thoroughly cleaned and relubricated daily.

(2) If a considerable amount of dust or sand is present, the teeth should be wiped dry before the gun is operated, situation permitting, and then be relubricated after the action is over. With the surfaces dry, there is less wear than when they are coated with lubricant contaminated with sand.

(3) Check the elevating mechanism for evidences of backlash. Report to the ordnance maintenance personnel for correction. Backlash will result in excessive muzzle whip and inaccuracies in firing.

d. Traversing Mechanism. Check the traversing mechanism for evidences of backlash. Report to the ordnance maintenance personnel for correction. Backlash will result in incorrect indicator readings and inaccuracies in firing.

e. Front and Rear Chassis Compensating Spring Lock and Cover.

(1) When engaging the front and rear chassis compensating spring locks in the toothed chassis locking arcs when lowering and raising the weapon, care should be taken to insure that the lock is fully engaged and alined with one of the notches in the arc. Excess pressure on the lock handle when teeth are not engaged properly will result in the shearing of the taper pins which secure the handle to the lock spindle.

(2) In cold weather operations, water and moisture seeping into the housing of the axle lock will, upon freezing, prevent operation of the lock. The lock should be kept well lubricated with OIL, engine, SAE 10, or OIL, lubricating, preservative, light. The top of the lock shank and housing should be heavily covered with grease to prevent water from seeping into the housing. The opening between the housing and the lock should be greased each time the carriage is raised or lowered, as the movement of the lock will break the grease seal. An improvised cap may be made to prevent moisture from collecting in the housing.

(3) All carriages are required to be modified by the addition of chassis clips and web straps, to prevent the front and rear chassis compensating spring locks from becoming accidentally disengaged from the chassis locking arcs. These straps should secure the lock handles in locked position at all times, except when the lock handle is used to unlock the axles.

TM 9-252

## CARE AND PRESERVATION

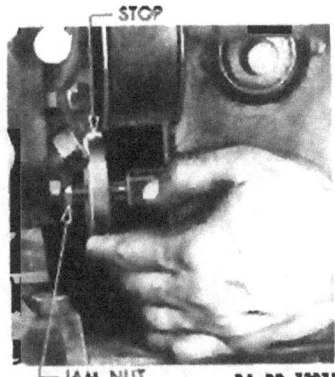

*Figure 113 — Front Firing Pedal — Proper Amount of Travel*   *Figure 114 — Front Firing Pedal Stop and Nut — Adjustment*

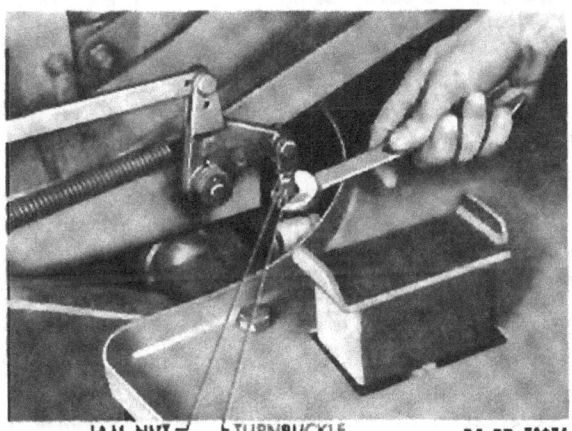

*Figure 115 — Rear Firing Pedal — Adjustment*

**f. Firing Mechanism.**

(1) TO ADJUST FRONT FIRING PEDAL. For sufficient movement for proper functioning, the front firing pedal when released should be 3½ inches from the inner footrest (fig. 113). To adjust, loosen the jam nut on the inner front firing pedal link and rotate the front pedal link firing stop (fig. 114). When adjustment has been made, tighten the jam nut.

(2) TO ADJUST REAR FIRING PEDAL. The rear firing pedal should have no free travel. To adjust, loosen the jam nut on the firing mechanism turnbuckle. Turn the turnbuckle up (fig. 115) until

147

## 40-MM AUTOMATIC GUN M1 (AA) AND 40-MM ANTIAIRCRAFT GUN CARRIAGES M2 AND M2A1

*Figure 116 — Breech Casing Firing Mechanism — Assembled*

*Figure 117 — Breech Casing Firing Mechanism — Adjustment*

free travel is felt in depressing the pedal. Then turn the turnbuckle down until the free travel of the pedal is eliminated. Tighten the jam nut.

(3) TO ADJUST BREECH CASING FIRING MECHANISM FOR SINGLE FIRE. If gun will not single fire, adjust the clearance between the firing lever pawl and the trigger. The pawl pivots on an axis pin mounted through lugs on the firing lever (fig. 117). To adjust, disconnect the front firing rod link from the firing crank by removing the cotter pin and axis pin. Loosen the firing rod link lock nut. Screw the firing rod link in or out to adjust the clearance. Tighten the lock nut. Connect the firing rod link to the firing crank.

g. Level Assemblies. Check the level assemblies on the base plate of the top carriage. Place the weapon in firing position on a flat surface. Level the weapon by means of the level assemblies. Traverse the weapon 180 degrees, check the levels, and relevel as necessary. Traverse the weapon 360 degrees, checking the levels at each 90 degrees. If the levels do not indicate levelness during the full traverse of the weapon, check the levelness of the weapon in full traverse with a gunner's quadrant placed on the machined section of the breech casing. If the levels are at fault, or if the weapon will not remain level in a full traverse, report the matter to the ordnance maintenance personnel.

h. Pneumatic Tires and Tubes.

(1) The pressure to be maintained in the tires is 45 pounds per

## CARE AND PRESERVATION

square inch. To obtain maximum mileage, the air pressure in the tires should be checked prior to operations and thereafter left alone unless there is a loss of air pressure. Bleeding of air from the tires results in an increase of the flexing of the tire side walls which increases the danger of failure.

(2) Remove all foreign substance from the rubber, being especially careful to keep tires as free from oil and grease as possible. Oil and grease have a deteriorating effect upon rubber.

i. Care of Electric Brakes. Proper attention must be given the electrical brake system to insure proper functioning. It is essential that the using arm personnel has knowledge which will permit the detection and correction of malfunctions which may occur during traveling. This knowledge will also permit the best possible service from this equipment. The following information and that contained in paragraph 57 a through d will prove helpful:

(1) WIRING AND CONTROLLER. Check the wiring and the controller before examining the brakes. Check for poor connections and partly broken or worn wires. Test the battery of the prime mover to see that it is sufficiently charged to turn over the starting motor. When brakes are new, several applications must be made before maximum efficiency is obtained.

(2) CHECK CURRENT AT ELECTRIC BRAKES.

(a) Check current at the brakes, using the ammeter furnished as an accessory. Disconnect one brake wire only. Connect one side of the ammeter to the brake, the other side to the terminal of the live wire that was removed from the brake. Leave the other brakes in the circuit. Take a reading; it should not be less than 2.2 amperes. If this amount of current is not flowing through each brake, proper operation will not be obtained.

(b) Check current consumption of each of the brakes, one at a time, leaving the others in circuit. The readings should not vary more than 0.5 ampere. In case there are greater variations, check all connections for poor contacts or broken wires. As the tests are completed, remove the ammeter from the circuit and connect the live wires to the brake terminals.

(3) CHECK DRY CELL BATTERY. Check the battery used to actuate the electric brakes when the carriage breaks away from the prime mover. The using arm personnel must make certain that this battery will produce sufficient current to actuate the brakes at all times. All wiring and switches must function correctly.

(4) FACING OF MAGNET. The facing of the magnet may become glazed. This is due to some foreign substance embedding itself into the material, resulting in a polished surface. If the facing cannot be roughened with coarse CLOTH, abrasive, aluminum-oxide, notify ordnance maintenance personnel.

(5) STOP LIGHTS. Stop lights must not be connected into the

TM 9-252
67

## 40-MM AUTOMATIC GUN M1 (AA) AND 40-MM ANTIAIRCRAFT GUN CARRIAGES M2 AND M2A1

brake circuit. This changes the amount of current which passes through the controller, resulting in weak or grabbing brakes.

(6) BEARINGS AND WHEELS. Worn bearings or loose wheels will cause erratic action of the brakes and will be evidenced by the wide track the pole faces of the magnets make on the armatures. The roller bearings must be adjusted (par. 88 d); broken or badly worn bearings must be replaced. Notify ordnance maintenance personnel.

(7) Notify ordnance maintenance personnel when brake lining is worn out or has become greasy, brake bands have become distorted, brake drums are out of round, or other conditions are found, the corrective measures for which are beyond the scope of using arm personnel.

j. Adjustment of Brakes.

(1) Remove wheel, hub, and drum assemblies, and clean out accumulated foreign substance. See that shoes move freely on backing plate and are seated against the anchor pin.

(2) Reinstall hub and drum assemblies and loosen nut on secondary shoe equalizer just enough to allow equalizer to turn.

(3) Turn the equalizer in direction the wheel revolves when in forward motion until brake has considerable drag.

(4) Turn brake drum so that slot for feeler gage is near equalizer and insert 0.010-inch feeler gage between drum and brake lining. Then turn the equalizer in opposite direction very slowly until only a slight drag is felt on feeler gage. Tighten the equalizer, lock nut, and recheck clearance to see that equalizer has not moved while tightening lock nut. Use a $3/16$-inch wrench on brake equalizer and a $9/16$-inch wrench on lock nut.

(5) With the secondary shoe properly adjusted, remove the small metal cover in the backing plate and expand the adjusting screw until brake drags slightly. Then release the adjusting screw until only a slight drag is felt on a 0.010-inch feeler gage, placed at the center of the primary shoe.

NOTE: Be sure to replace brake backing plate adjusting hole covers to prevent entry of foreign substances.

k. Lighting.

(1) The lighting fixtures on the rear of the carriage are equipped for both service lighting and blackout lighting. The change-over from service to blackout lighting, and from blackout to service lighting is made by lifting the small cover on the right side of the rear end of the chassis and turning the slot to indicate "S" (service) or "BO" (blackout).

(2) STOP AND TAILLIGHT LAMP REPLACEMENT. Remove the two lamp cover retaining screws. Remove the cover. Lamp units can be withdrawn straight back out of their sockets in the lamp housing.

## CARE AND PRESERVATION

1. **Care of Canvas.** To prevent formation of damaging mildew during periods of storage, shape out and air the canvas cover for several hours at frequent intervals. Repair without delay any loose grommets or rips in the canvas; failure to make immediate repairs may allow a minor defect to develop into major damage. Mildewed canvas is best cleaned by scrubbing with a dry brush. If water is necessary to remove dirt, it must not be used until mildew has been removed. If mildew has been present, examine fabric carefully by stretching and pulling for evidence of rotting or weakening of fabric where mildew has been. If fabric shows indication of loss of tensile strength, it is probably not worth retreatment. Oil and grease can be removed by scrubbing with issue soap and warm water. Rinse well with clear water, and dry.

CAUTION: At no time is gasoline, or solvent to be used to remove oil or grease spots.

## 68. CLEANERS AND PRESERVING MATERIALS.

a. The following cleaners, abrasives, and preservatives are for use with this materiel. See TM 9-850 for detailed information.

(1) CLEANERS AND ABRASIVES.

ALCOHOL, ethyl
BURLAP, jute (8-oz 40 in. wide)
CLEANER, rifle bore
CLOTH, abrasive, aluminum oxide
CLOTH, crocus
CLOTH, wiping
LIME, hydrated
PAPER, flint, class B, Nos. 00, 1, 2
PAPER, lens, tissue
PATCHES, cut (canton flannel)
REMOVER, paint, and varnish
SOAP, castile
SOAP, liquid, lens cleaning
SODA, ASH
SODA, caustic, lye
SOLVENT, dry-cleaning
SPONGE, natural or cellulose
WASTE, cotton, white

(2) LUBRICATING AND PRESERVING MATERIALS.

COMPOUND, rust-preventive, heavy
COMPOUND, rust-preventive, light
GREASE, graphited, light
GREASE, lubricating, special
GREASE, O.D., No. 0
GREASE, O.D., No. 00
OIL, engine, SAE 10
OIL, engine, SAE 30
OIL, hydraulic
OIL, lubricating, for aircraft instruments and machine guns
OIL, lubricating, preservative, light
OIL, lubricating, preservative, medium
OIL, recoil, light
OIL, recoil, special
PETROLATUM
SOAP, saddle

## 69. PAINTS AND RELATED MATERIALS.

a. The following paints and related materials are required for use

## 40-MM AUTOMATIC GUN M1 (AA) AND 40-MM ANTIAIRCRAFT GUN CARRIAGES M2 AND M2A1

with this materiel. See Table of Allowances and TM 9-850 for detailed information.

ENAMEL, synthetic, gloss-red
ENAMEL, synthetic, olive-drab lusterless
ENAMEL, synthetic, stenciling, lusterless
NAPHTHALENE, balls
OIL, neat's-foot
PRIMER, synthetic, refinishing
PRIMER, synthetic, rust-inhibiting
THINNER, for synthetic enamels
THINNER, paint, volatile mineral spirits
VARNISH, shellac

### 70. MISCELLANEOUS MATERIALS AND TOOLS.

a. The following miscellaneous materials and tools are required for use with this materiel. See Table of Allowances and TM 9-850 for detailed information.

BRUSH, artist, camel's-hair, rd., No. 1
BRUSH, flowing, skunk's-hair, No. 3 (2-in.)
BRUSH, paint, metal bound, flat
BRUSH, sash-tool, oval, No. 1
BRUSH, scratch, wire, painter's, (14-in. x ⅞-in.) handled
BRUSH, varnish, oval (1⅞-in.)
CHALK, white, railroad, 1-in. x 4-in.
KNIFE, putty, 1¼-in. type IV
NEEDLES, sacking
PALM, sailmaker's
TWINE, jute

b. The bristles of brushes are subject to attack by moths. Brushes in storage should be protected by NAPHTHALENE balls.

c. Brushes used for painting after being thoroughly cleaned with THINNER, paint, volatile mineral spirits should be laid flat on a horizontal surface (not in water).

### 71. WASHING.

a. Serious damage to ordnance materiel, in many cases requiring repair and replacement of component parts of sighting equipment, fire control instruments, weapons, and carriages, has frequently resulted from the use of water, steam, or air from a high-pressure hose for cleaning purposes. For this reason, operating personnel is cautioned to prevent water, dirt, or grit from being forced into any part of the gun or carriage when using water, steam, or air under pressure for cleaning.

b. Under no circumstances will a hose, either normal-pressure or high-pressure, be used in cleaning any sighting equipment or any fire control instruments. Before washing, take off removable sighting equipment from the materiel to be cleaned. In cases where it is not removable, take care to cover the parts properly.

## CARE AND PRESERVATION

### 72. PAINTING.

a. General.

(1) Ordnance materiel is painted before being issued to the using arms and one maintenance coat per year will ordinarily be ample for protection. With few exceptions, this materiel will be painted with ENAMEL, synthetic, olive-drab, lusterless. The enamel may be applied over old coats of long-oil enamel and oil paint previously issued by the Ordnance Department if the old coat is in satisfactory condition for repainting.

(2) Paints and enamels are usually issued ready for use and are applied by brush or spray. They may be brushed on satisfactorily when used unthinned in the original package consistency, or when thinned no more than 5 percent by volume with THINNER, for synthetic enamels. The enamel will spray satisfactorily when thinned with 15 percent by volume of this thinner. (Linseed oil must not be used as a thinner since it will impart a luster not desired in this enamel.) If sprayed, it dries hard enough for repainting within ½ hour and dries hard in 16 hours.

(3) Certain exceptions to the regulations concerning painting exist. Fire control instruments, for instance, which require a crystalline finish, will not be painted by the using arms.

(4) Complete information on painting is contained in TM 9-850.

b. Preparation for Painting.

(1) If the base coat on the materiel is in poor condition and it is desirable to strip the old paint from the surface rather than to use sanding and touch-up methods, it will be necessary to apply a primer coat after the old finish has been removed.

(2) PRIMER, synthetic, refinishing, should be used on wood as a base coat for synthetic enamel or as a second coat over PRIMER, synthetic, rust-inhibiting. It may be applied either by brushing or spraying. It will brush satisfactorily as received, or after the addition of not more than 5 percent by volume of THINNER, for synthetic enamels. It will be dry enough to touch in 30 minutes, and hard in 5 to 7 hours. For spraying, it may be thinned with not more than 15 percent by volume of THINNER. Lacquers must not be applied to the PRIMER, synthetic, refinishing, within less than 48 hours.

(3) PRIMER, synthetic, rust-inhibiting, for bare metal, should be used on metal as a base coat. Its use and application are similar to that outlined in step (2), above.

(4) The success of a job of painting depends partly on the selection of a suitable paint, but also largely upon the care used in preparing the surface prior to painting. All parts to be painted should be free from rust, dirt, grease, kerosene, and alkali, and must be dry.

c. Painting Metal Surfaces.

(1) Metal parts may be washed in a liquid solution, consisting of ½ pound of SODA ASH in 8 quarts of warm water, then rinsed

**TM 9-252**

### 40-MM AUTOMATIC GUN M1 (AA) AND 40-MM ANTIAIRCRAFT GUN CARRIAGES M2 AND M2A1

in clean water and wiped thoroughly dry. Wood parts may be treated in the same manner but the alkaline solution must not be left on for more than a few minutes and the surfaces should be wiped dry as soon as they are washed clean.

(2) When artillery is in fair condition and only marred in spots, the bad places should be touched with ENAMEL, synthetic, olive-drab, lusterless, and permitted to dry. The whole surface should then be sandpapered with PAPER, flint, class B, No. 1, and a finish coat of ENAMEL, synthetic, olive-drab, lusterless, applied and allowed to dry thoroughly before the materiel is used.

(3) If the equipment is in bad condition, all parts should be thoroughly sanded with PAPER, flint, class B, No. 2, given a coat of PRIMER, synthetic, refinishing, and permitted to dry for at least 16 hours. Sandpaper with PAPER, flint, class B, No. 00, wipe free from dust and dirt, and apply a final coat of ENAMEL, synthetic, olive-drab, lusterless. Allow the materiel to dry thoroughly before it is used.

*d.* Paint as a Camouflage. Camouflage is now the major consideration in painting ordnance materiel, with rust prevention secondary. The camouflage plan at present employed utilizes color and freedom from gloss.

(1) COLOR. Materiel is painted with ENAMEL, synthetic, olive-drab, lusterless, which was chosen to blend in reasonably well with the average landscape.

(2) GLOSS. The new lusterless enamel makes the materiel difficult to see from the air or from great distances over land. Materiel painted with ordinary glossy paint can be detected more easily and at greater distances.

(3) PRESERVING CAMOUFLAGE.

*(a)* Since continued friction or rubbing will smooth the surface and produce a gloss, it must be avoided. The materiel should not be washed more than once a week. Care should be taken to see that the washing is done entirely with a sponge or a wiping cloth. The surface should never be rubbed or wiped, except while wet, or a gloss will be developed.

*(b)* It is not desirable that materiel painted with lusterless enamel be kept as clean as when glossy paint was used. A small amount of dust increases the camouflage value. Grease spots should be removed with SOLVENT, dry-cleaning. Whatever portion of the spot cannot be so removed should be allowed to remain.

*(c)* Continued friction of wax-treated tarpaulins on the sides of the materiel will also produce a gloss. It should be removed with SOLVENT, dry-cleaning.

*(d)* Tests indicate that repainting will be necessary once yearly

in the case of the olive-drab and twice yearly in the case of the blue-drab enamel.

e. Removing Paint.

(1) After repeated paintings, the paint may become so thick as to scale off in places and present an unsightly appearance. If such is the case, remove the old paint by use of a lime-and-lye solution or REMOVER, paint and varnish. It is important that every trace of lye or other paint remover be completely rinsed off and that the equipment be perfectly dry before repainting is attempted. It is preferable that the use of lye solutions be limited to iron or steel parts.

(2) If used on wood, the lye solution must not be allowed to remain on the surface for more than a minute before being thoroughly rinsed off and the surface wiped dry with wiping cloth or waste. Crevices or cracks in wood should be filled with putty and the wood sandpapered before finishing. The surfaces thus prepared should be painted according to the directions given previously.

f. Painting Lubricating Devices. Oil cups, grease gun fittings, oilholes, and similar lubricating devices, as well as a circle about ¼ inch in diameter at each point of lubrication will be painted with ENAMEL, synthetic, gloss-red, in order that they may be readily located.

NOTE: Do not paint the fittings themselves.

## Section VIII

## INSPECTION AND ADJUSTMENT

### 73. PURPOSE.

a. Inspection of your weapon is vital. Thorough, systematic inspection at regular intervals is the best insurance against an unexpected gun breakdown at the critical moment when maximum performance is absolutely necessary. Never let your materiel run down, keep it in first class fighting condition by vigilant inspection and prompt maintenance.

b. Inspection is for the purpose of determining the condition of the materiel, whether repairs or adjustments are required, and the remedies necessary to insure serviceability and proper functioning. Its immediate aim is trouble prevention, which includes:

(1) Preventive maintenance.

(2) Discovering evidence of improper treatment received by the materiel before delivery into your hands.

TM 9-252
73-75

**40-MM AUTOMATIC GUN M1 (AA) AND 40-MM ANTIAIRCRAFT GUN CARRIAGES M2 AND M2A1**

*Figure 118 — Tube Serial Number*

(3) Determining when replacements of parts is necessary because of ordinary wear or defects in parts.

## 74. VISUAL INSPECTION UPON RECEIPT.

a. Upon receipt of this materiel, it is the responsibility of the officer in charge to ascertain whether the materiel is complete and in sound operating condition. A record should be made of any missing parts and of any malfunctions, and any such conditions should be corrected as quickly as possible.

b. Attention should be given to small and minor parts, as these are the more likely to become lost and their lack may seriously affect the proper functioning of the materiel.

c. This visual inspection upon receipt should be followed as quickly as possible by a complete inspection which will disclose the functioning of the materiel (pars. 76 and 77).

## 75. SERIAL NUMBERS.

a. Three serial numbers are required for records concerning the components of this materiel. They are the tube serial number, the gun serial number, and the carriage serial number.

b. **Tube Serial Number.** The tube serial number is stamped in the top surface of the breech end of the tube (fig. 118).

c. **Gun Serial Number.** The gun serial number is stamped on a metal plate mounted on top of the breech casing directly in front of the automatic loader (fig. 119).

156

TM 9-252
75-76

INSPECTION AND ADJUSTMENT

Figure 119 — Gun Serial Number

Figure 120 — Carriage Serial Number

d. **Carriage Serial Number.** The carriage serial number is stamped on a metal plate mounted on the elevating mechanism worm wheel cover on the left side of the carriage (fig. 120).

## 76. INSPECTION OF GUN.

a. The following instructions will be carefully observed by all concerned:

| Parts To Be Inspected In Order of Inspection | Points To Observe |
|---|---|
| Gun as a unit. | Note general appearance. Test smoothness of operation of breech mechanism. Test operation of firing mechanism, first by cocking the gun and depressing rear firing pedal; then by cocking the gun and depressing forward firing pedal. Open and close top cover, side cover, and bottom cover; note fit and security with which they lock. |
| Barrel assembly. | Remove barrel assembly. Examine interrupted threads and recesses in breech face for scoring or other mutilation. Note the condition of the bore for wear on lands or deposits in grooves; erosion at origin of rifling. Examine barrel guide sleeve for burs, wear, or mutilations. See that barrel guide sleeve and |

TM 9-252
76-77

**40-MM AUTOMATIC GUN M1 (AA) AND 40-MM ANTIAIRCRAFT GUN CARRIAGES M2 AND M2A1**

| Parts To Be Inspected In Order of Inspection | Points To Observe |
|---|---|
| | locking collar are properly positioned and that barrel guide sleeve locking set screw and friction disk are tightly screwed into place. Note condition of copper gasket at rear of flash hider. Examine condition of recuperator spring. Replace barrel assembly. |
| Breechblock. | Remove breechblock, inner cranks, and extractors. Inspect for wear, burs, or other mutilations. Note condition of impact surfaces on front face of breechblock. See that parts are clean and well lubricated. Be sure breechblock firing pin bushing locking pin is secure and flush with front face of breechblock. Measure protrusion of firing pin with striker protrusion gage. Replace extractors, breechblock, and inner cranks. |
| Recoil cylinder. | Check for proper amount of recoil liquid. Note recoil cylinder anchor bracket for its proper assembly (par. 86 f). Check setting of counterrecoil adjusting valve. Inspect for leaks. |
| Automatic loader. | Insert a clip of dummy ammunition in automatic loader and check for proper operation. |

## 77. INSPECTION OF CARRIAGE.

a. The following instructions will be carefully observed by all concerned:

| | |
|---|---|
| Carriage. | Note general appearance. Examine all covers. Note whether oil and grease fittings are clean and painted red, that a red ring has been painted around all oilholes, and that carriage is painted in accordance with regulations. |

## INSPECTION AND ADJUSTMENT

| Parts To Be Inspected in Order of Inspection | Points To Observe |
|---|---|
| Compensating springs. | Lower carriage to ground. Examine action and effectiveness of locking handles and compensating springs. Test levelness of carriage in traverse (par. 67 g). While wheels are off ground, rotate to test for drag. After inspecting elevating and traversing mechanism, raise carriage. |
| Elevating mechanism. | Elevate and depress gun manually throughout full extent of travel. Note whether same effort is required to elevate and depress. Examine for excessive backlash. Check elevating arc for dirt, rust, wear, burs, and protective lubrication. Check lubrication in gear case. Be sure there is no leakage. Test elevation by remote control. Check oil gears for oil levels in oil gear, chain case, and clutch housing; creep and dither. Creep should be as near zero as possible; dither should be noticeable to touch. |
| Traversing mechanism. | Traverse by hand, left and right, 360 degrees to check smoothness of action. Examine for excessive backlash. Check lubrication in gear cases. Be sure there is no leakage. Test traverse by remote control. Check oil gears for oil levels, creep, and dither. Test operation of power synchronizing (slewing) handle. |
| Top carriage and carriage frame. | Examine for breaks and cracks, and for loose attachment screws and nuts. Determine whether outrigger hinges and leveling jacks are lubricated adequately and operate and lock properly. |

TM 9-252
77-78

**40-MM AUTOMATIC GUN M1 (AA) AND 40-MM ANTIAIRCRAFT GUN CARRIAGES M2 AND M2A1**

| Parts To Be Inspected in Order of Inspection | Points To Observe |
|---|---|
|  | Examine stake and other accessory mountings on frame, outriggers, and loading platform. See that frame and swivel body have been drained. Inspect lower surfaces of carriage for rust. |
| Drawbar and gun stay. | Look for undue wear on lunette. Examine drawbar connections and pins. Check connections and operation of upper plungers of gun stay. |
| Brakes. | Tow carriage and test effectiveness of brakes by making several stops. Test action of hand brakes. Travel slowly and pull safety switch chain; carriage should stop suddenly. |
| Wheels and tires. | Check for any loose or missing nuts. Check tires for correct pressure, 45 pounds per square inch. Note condition of tires and whether treads are taking wear evenly. |
| Lights. | Test service stop lights and taillights. Turn blackout light switch and test blackout stop lights and taillights. Check condition of reflectors. |

## Section IX

## DISASSEMBLY AND ASSEMBLY

**78. GENERAL.**

a. Wear, breakage, cleaning, and inspection make necessary the occasional disassembly and assembly of various parts of the weapon. This work comes under two headings, that which can be performed by the using arm personnel, and that which must be performed by ordnance maintenance personnel.

b. The using arm personnel may, in general, do such dismounting,

## DISASSEMBLY AND ASSEMBLY

replacing, repairing, and installing of parts as may be required for cleaning, lubrication, inspection, adjustment, and the installation of such spare parts as are regularly issued to the using arms, or which may be installed with the tools and equipment available to the using arm personnel. All work should be done in the manner prescribed and with the proper tools. Any difficulty which cannot be overcome must be brought to the attention of ordnance maintenance personnel.

c. The using arm personnel will not attempt unauthorized disassembly of any unit of the weapon nor any filing of parts other than specified in this manual, except by order of the commanding officer.

d. Avoid the use of wrenches that do not fit snugly on the parts or of screwdrivers that do not fit the slots in screws properly. The poor fitting of these tools will not only damage the corners of nuts and bolt heads and the slots of screws, but may also damage the tools.

e. Before attempting to put together the larger assemblies which compose the weapon, the assembly of subassemblies should be completed. In this work, all bearings, sliding surfaces, threads, and enclosed portions of parts should be cleaned and oiled for lubrication as well as for the prevention of rust. Care must be exercised to exclude dust, moisture, and other foreign matter.

f. A steel hammer, chisel, or screwdriver should not be used to drive directly against any part of the gun or machined part of the carriage. A copper or wooden mallet, or a copper-, rawhide-, or plastic-faced hammer should be used, or a brass or copper drift or a hardwood block should be interposed between the blow and the part to be struck.

## 79. BARREL ASSEMBLY REMOVAL AND INSTALLATION.

a. The removal and replacement of the barrel assembly is a using arms operation. Barrel assemblies of the 40-mm Automatic Gun M1 are interchangeable in the breech ring assemblies of all 40-mm Automatic Guns M1.

b. The barrel assembly may be removed independently, or it may be removed as part of a sequence of operations including removal of the automatic loader and breech ring. If the automatic loader and breech ring are to be removed, a slightly different series of operations must be performed than when the barrel assembly is to be removed independently.

c. Barrel Assembly Removal (When Breech Ring Is Not To Be Removed).

(1) Place the gun in horizontal position. Secure the breech casing to the gun stay. Open the breech by means of the hand operating lever. Insert the breechblock locking pin in its appropriate holes in the breech casing and breech ring (fig. 121).

## 40-MM AUTOMATIC GUN M1 (AA) AND 40-MM ANTIAIRCRAFT GUN CARRIAGES M2 AND M2A1

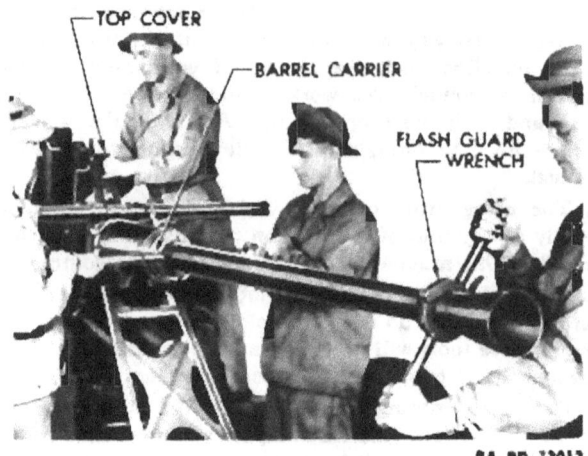

*Figure 121 — Barrel Assembly — Unlocking*

(2) Assemble the flash guard wrench on its seat on the muzzle end of the tube. Be sure that the hinged portion of the wrench is placed at the top. When assembled in this position, the hinged portion of the wrench will be beneath the tube after the tube has been rotated one-half turn, and there will be less likelihood of the wrench being broken when the tube is being carried.

(3) Place the barrel carrier under the tube, fitting the lug on the carrier in the groove on the barrel guide sleeve locking collar (fig. 121). Place one man on each handle of the barrel carrier, as these men must support the greater portion of the weight of the barrel assembly when it is removed from the breech casing. One man can handle the flash guard wrench and be in a better position to maneuver it properly than if two men manipulate it.

CAUTION: In carrying the barrel assembly, keep it on an even keel. If the muzzle end is raised, there is a possibility that the barrel assembly will become detached from the barrel carrier, permitting the breech of the barrel to strike the ground.

(4) Open the top cover on the breech casing by lifting the cover latch lever head and rotating the cover latch lever (fig. 122) clockwise, depressing the top cover catch, and lifting the top cover by its handle. The cover will lock automatically in upright position. Opening the top cover will lift and release the breech ring barrel catch.

(5) Rotate the tube one-half turn counterclockwise (fig. 121). This will unlock the interrupted threads holding the tube in the breech ring.

TM 9-252
79

## DISASSEMBLY AND ASSEMBLY

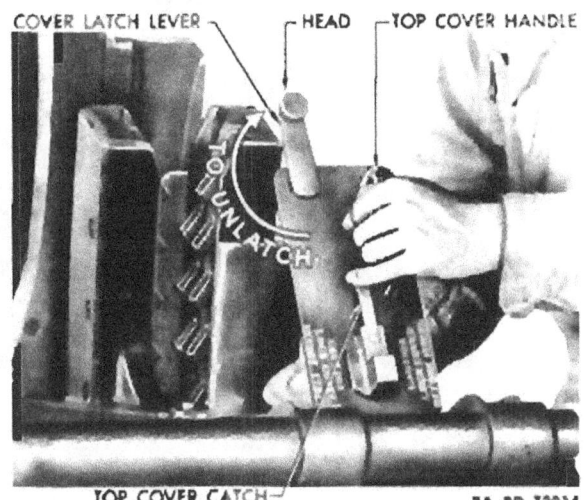

*Figure 122 — Top Cover — Opened*

*Figure 123 — Barrel Assembly — Removal*

(6) Withdraw the barrel assembly straight forward from the breech casing (fig. 123), with the men supporting the barrel carrier exercising caution that the interrupted threads on the tube do not come in contact with the breech casing as the barrel guide sleeve slides from the casing.

CAUTION: Do not attempt to remove the barrel assembly unless

163

TM 9-252
79-80

**40-MM AUTOMATIC GUN M1 (AA) AND 40-MM ANTIAIRCRAFT
GUN CARRIAGES M2 AND M2A1**

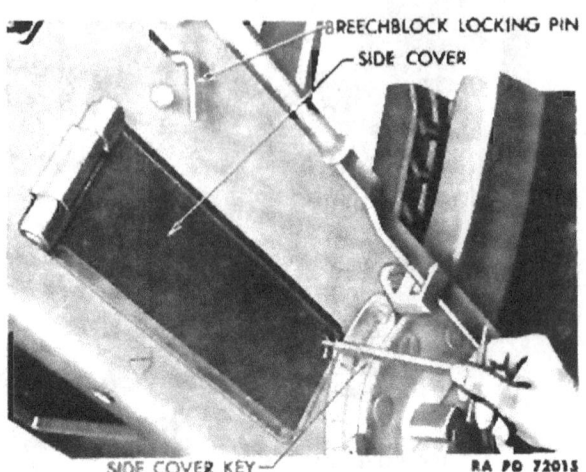

*Figure 124 — Side Cover — Opening With Key*

the breech is open or the breechblock has been removed from the breech ring. To do so will damage the extractors and produce burs in the extractor grooves in the breech end of the barrel.

d. **Barrel Assembly Installation (When Breech Ring Has Not Been Removed).**

(1) Check to see that the top cover on the breech casing is in the open position. Check to see that breechblock is open.

(2) Assemble the flash guard wrench on its seat with the hinged portion of the wrench under the tube. Place the barrel carrier under the tube, catching its lug in the groove of the collar.

(3) Lift the barrel assembly and insert the breech end of the tube in the front opening of the breech casing, carefully sliding it into the casing to prevent damage to the threads. By means of the flash guard wrench, rotate the tube one-half turn clockwise.

(4) Release the top cover catch, close the top cover, and lock it with the cover latch lever, locking the barrel assembly in the breech ring with the breech ring barrel catch. Remove the wrench and carrier.

**80. BREECHBLOCK, CLOSING SPRING, AND EXTRACTOR REMOVAL AND INSTALLATION.**

NOTE: The breechblock and closing spring assemblies may be removed without dismantling other parts of the weapon. If, however, the breech ring is to be removed, the breechblock and closing spring

## DISASSEMBLY AND ASSEMBLY

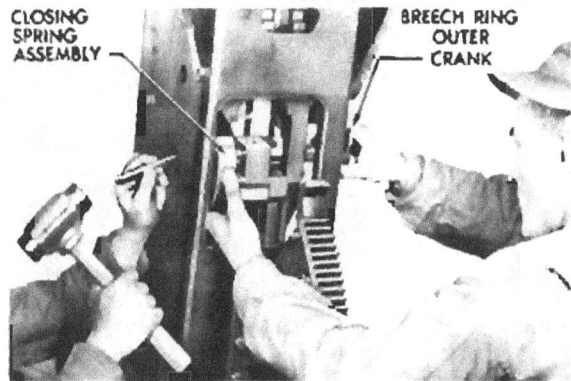

*Figure 125 — Closing Spring Assembly Removal*

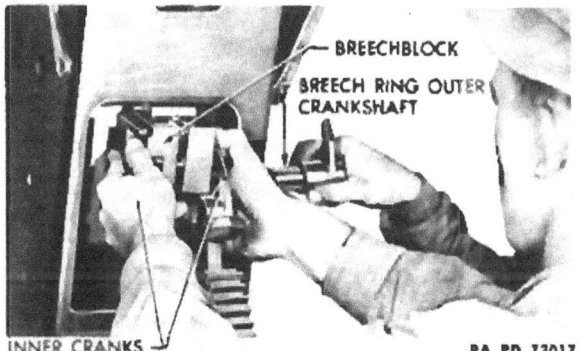

*Figure 126 — Breechblock and Inner Cranks Removal*

assemblies must be removed before the barrel assembly is taken from the weapon.

a. **Breechblock and Closing Spring Assemblies Removal.**

(1) Open the side cover by inserting the side cover key (fig. 124), pressing the key inwardly, and turning it counterclockwise. Tie back the side cover with a piece of cord to retain it in open position.

(2) Close the breech. Elevate the gun to about 45 degrees. Release the bottom cover by rotating the cover latch lever and remove this cover by disengaging its rear end from the breech casing.

(3) Through the smaller hole in the right side of the breech casing, tap the breech ring outer crankshaft to the left about 2 inches, sufficient to permit the removal of the breech ring closing spring assembly. Use a brass drift. Remove the closing spring assembly (fig. 125).

TM 9-252
80

**40-MM AUTOMATIC GUN M1 (AA) AND 40-MM ANTIAIRCRAFT GUN CARRIAGES M2 AND M2A1**

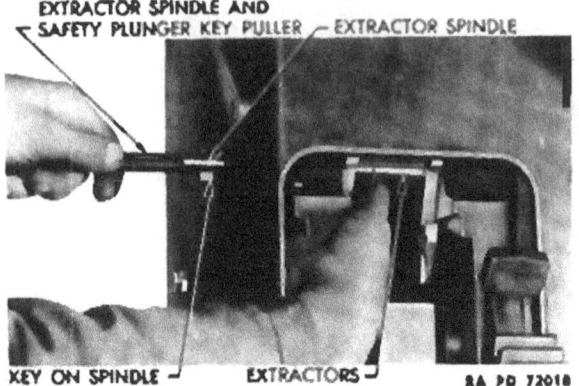

Figure 127 — Extractor Removal from Assembled Gun

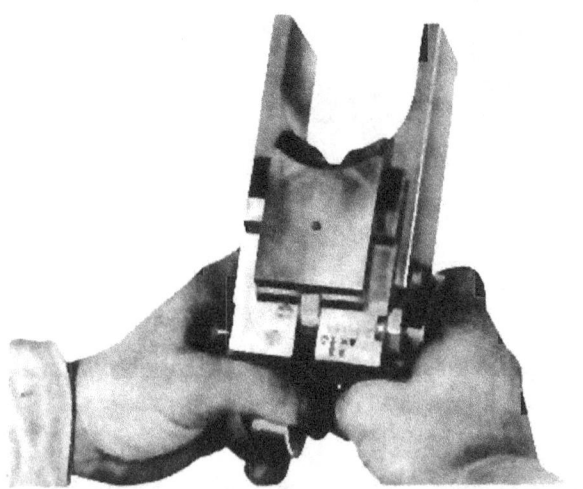

Figure 128 — Breechblock and Inner Cranks Held for Assembly

(4) Support the breechblock and right and left inner cranks, and pull the breech ring outer crank and crankshaft all the way out. Guide the breechblock assembly with the inner cranks out of the breech ring and through the opening in the bottom of the breech casing (fig. 126).

b. **Extractors Removal.**

NOTE: The cartridge extractors can be removed and replaced without removing the breech ring from the breech casing, or they can

## DISASSEMBLY AND ASSEMBLY

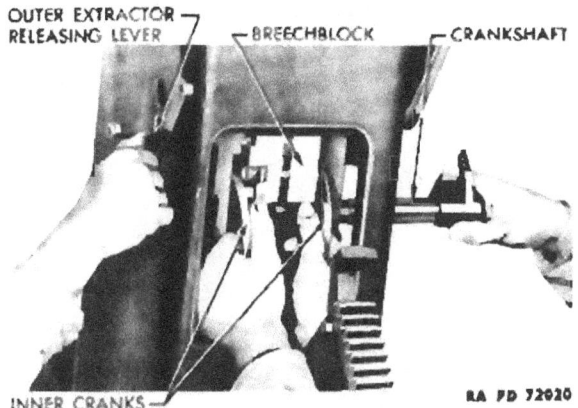

*Figure 129 — Breechblock and Inner Cranks — Installation*

be removed after the breech ring has been taken from the casing (par. 83 a). To remove the extractors from the assembled gun, the breechblock must first be removed (subpar. n, above).

(1) Through the larger of the two holes in the right side of the breech casing, screw the extractor spindle and safety plunger key puller into the threaded hole in the end of the extractor spindle.

(2) Reach through the bottom opening in the breech casing and support the extractors to keep them from falling out of the weapon after they have been released. Rotate the extractor spindle by means of the puller until the key on the spindle is mated with the keyway in the extractors. Withdraw the extractor spindle by means of the puller (fig. 127). Withdraw the extractors.

c. **Extractor Installation.** Screw the puller into the threaded end of the extractor spindle. Assemble the two extractors together. Place and hold the extractors in position in the breech ring. Insert the extractor spindle through the larger hole in the right side of the breech casing and through the breech ring and extractors, mating the key and keyway. Unscrew the puller.

d. **Breechblock and Closing Spring Assemblies Installation.**

(1) Insert the lugs of the left and right breech ring inner cranks in the grooves in the sides of the breechblock. Hold these parts in the manner illustrated in figure 128. Holding the parts in this manner will automatically cock the percussion mechanism.

(2) With the bottom cover removed, insert the breechblock and inner cranks into the bottom of the breech ring until further inward movement is stopped by the extractors. Press the extractor release lever (fig. 129) and push the breechblock into the breech ring as far

## 40-MM AUTOMATIC GUN M1 (AA) AND 40-MM ANTIAIRCRAFT GUN CARRIAGES M2 AND M2A1

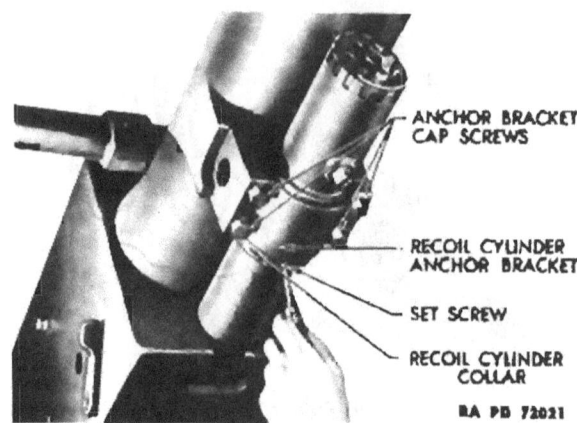

*Figure 130 — Recoil Cylinder Collar — Loosening Set Screw*

as it will go. Insert the breech ring outer crankshaft sufficiently to hold the breechblock in place.

(3) Insert the breech ring closing spring assembly and drive the crankshaft to its fully assembled position. Install the bottom cover.

## 81. FIELD STRIP DISASSEMBLY.

a. The field stripping of the gun involves the removal of the recoil cylinder, breech mechanism, barrel assembly, automatic loader and breech ring.

b. Before proceeding to disassemble the weapon by the series of operations required for the field strip, the breechblock must be in closed position and the rammer spring must be released. To release the rammer spring, move the outer safety lever from "SAFE" to "AUTO FIRE," move the feed control thumb lever to the right, and step on the firing pedal; then return the levers to their original positions.

c. Recoil Cylinder Removal.

NOTE: It is necessary that the recoil cylinder be removed to permit the removal of the breech ring from the breech casing.

(1) Elevate the gun to approximately 45 degrees. Loosen the set screw in the recoil cylinder collar (fig. 130). Back the collar from the recoil cylinder anchor bracket. If the recoil cylinder wrench is available, it should be used to remove the collar; otherwise, the collar should be loosened by being rotated with a brass drift and hammer.

(2) Support the cylinder and remove the four anchor bracket cap

## DISASSEMBLY AND ASSEMBLY

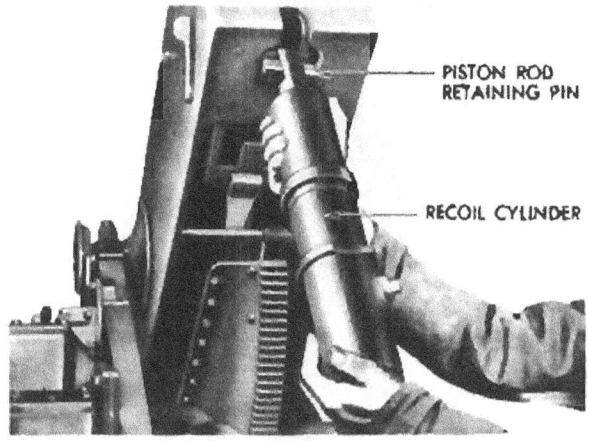

*Figure 131 — Recoil Cylinder — Removal*

screws. Remove the anchor bracket. Lower the recoil cylinder until it is at approximately right angles to the gun. Disengage the recoil cylinder piston rod retaining pin from the lugs of the breech ring (fig. 131), and remove the recoil cylinder.

d. **Breech Mechanism Removal.** Remove the breechblock, inner cranks, and closing spring case (subpar. a, above). The extractors need not be removed at this time. Detail strip the breechblock (par. 82). Disassemble the breechblock closing spring (par. 84).

e. **Barrel Assembly Removal.** Follow the instructions given in paragraph 79 c. The hand operating lever must not be moved as the breechblock has already been removed from the breech ring.

CAUTION: Be sure that the top cover remains in locked open position or that the breech ring locking pin is in its place in the holes in the breech casing and breech ring. These devices prevent the breech ring and automatic loading tray from moving to the rear after the barrel assembly has been removed.

f. **Rear Cover Release.** Remove the cartridge case deflector pin and press down the lower end of the deflector. Remove the nut from the rear cover attaching bolt. Drive out the attaching bolt (fig. 132). Lower the rear cover, being careful not to let it fall back violently as the bolt is removed.

g. **Automatic Loader Removal.**

NOTE: The automatic loader assembly may be removed independently of other parts of the weapon. The rear cover must first be lowered (subpar. f, above).

# TM 9-252
## 40-MM AUTOMATIC GUN M1 (AA) AND 40-MM ANTIAIRCRAFT GUN CARRIAGES M2 AND M2A1

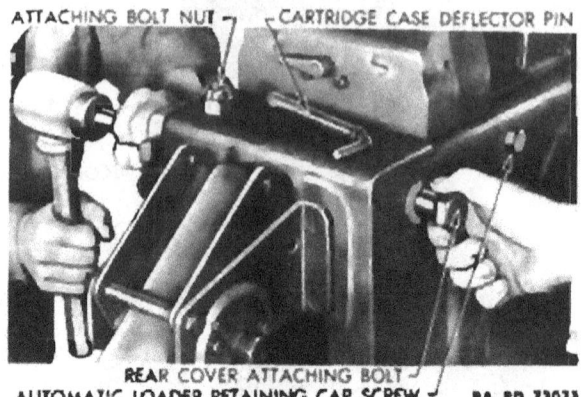

*Figure 132 — Rear Cover — Release*

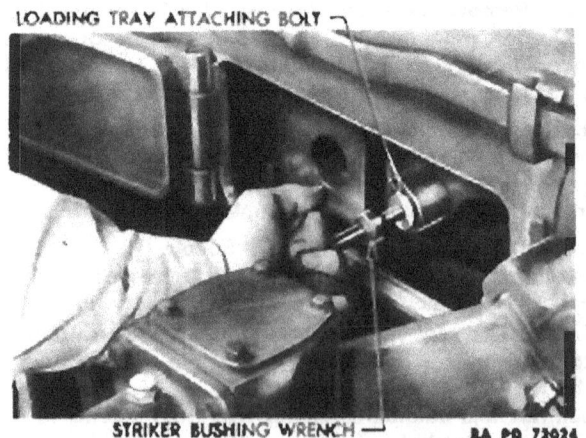

*Figure 133 — Loading Tray Attaching Bolt — Removal*

(1) Place the gun at zero elevation. Remove the automatic loader retaining cap screws from both sides of the breech casing (fig. 132).

(2) With the striker bushing wrench, remove the loading tray attaching bolt (fig. 133). To remove this bolt, it must be pressed inwardly with the wrench and turned 90 degrees in either direction.

(3) Be sure that the feed control thumb lever is in its left position.

TM 9-252
81

## DISASSEMBLY AND ASSEMBLY

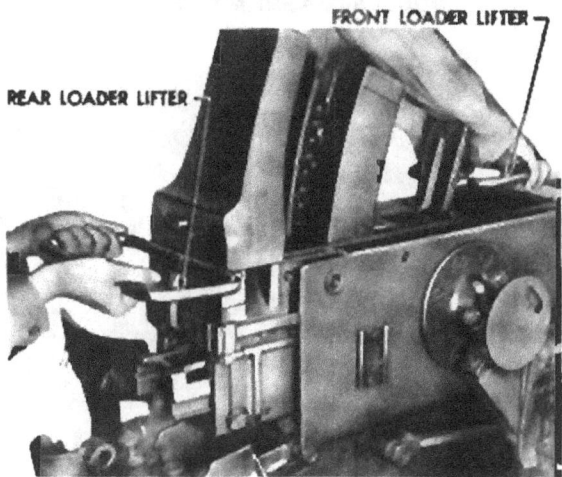

Figure 134 — Loader Lifters — Attachment

This is necessary in order to raise the right rammer check lever plunger and protect it from damage.

(4) Slide the loading tray backward to the rear edge of the breech casing to raise the center rammer catch lever plunger and protect it from damage when the automatic loader assembly is removed from the breech casing.

(5) Slide the entire automatic loader assembly approximately 5 inches to the rear to permit the attachment of the front and rear loader lifters. Attach the lifters (fig. 134).

(6) With a man grasping each of the loader lifters, lift the automatic loader to clear any obstruction, sliding it backward on its rails out of the breech casing through the opening made by the lowering of the rear cover.

CAUTION: When placing the automatic loader on a flat surface after its removal from the breech casing, great care must be taken to avoid damage to the feed roller catch release spindles. A slight tilting movement at the time of contact may result in bending the spindles.

(7) LOADING TRAY REMOVAL. After the automatic loader has been removed from the breech casing, slide the loading tray out the front end of the loader. Remove the loose loader feed rod rollers from the lower ends of the feed rods to prevent their being lost.

(8) LOADING TRAY INSTALLATION.

(a) Install the loader feed rod rollers on their pins in the lower ends of the feed rods. Insert the loading tray in the front end of the

TM 9-252
81

**40-MM AUTOMATIC GUN M1 (AA) AND 40-MM ANTIAIRCRAFT GUN CARRIAGES M2 AND M2A1**

*Figure 135 — Breech Ring — Removal*

loader and push it to the rear, making sure that the loader feed rod rollers are in their grooves in the sides of the loading tray. If, by accident or carelessness, the tray is inserted with the rollers on top of the grooves, the loader will be badly damaged when the weapon is fired.

(b) Do not push the loading tray too far to the rear of the loader. This will cause the catch and check levers to catch under the rammer shoe and the loading tray cannot be pulled forward for attachment to the breech ring.

(c) If the loading tray should be pushed too far, it may be disengaged in the following manner: Release the feed control check lever by rotating the feed control thumb lever. Release the firing mechanism check lever by rotating the rammer control spindle arm. Release the rammer catch lever by pressing the rammer release lever with a screwdriver and pulling the tray forward.

h. **Breech Ring Removal.**

NOTE: The recoil cylinder, breechblock and closing spring, barrel assembly, and automatic loader must be removed before the breech ring can be removed.

(1) Close the top cover to release the breech ring and the breech ring barrel catch. Remove the breechblock locking pin from its holes in the breech casing and breech ring, and place it in its bracket.

(2) Slide the breech ring assembly backward along its guideways and lift it out of the breech casing (fig. 135). Do not let the breech ring assembly drop as it comes to the end of the guideways.

(3) Detail strip the breech ring (par. 83).

172

## DISASSEMBLY AND ASSEMBLY

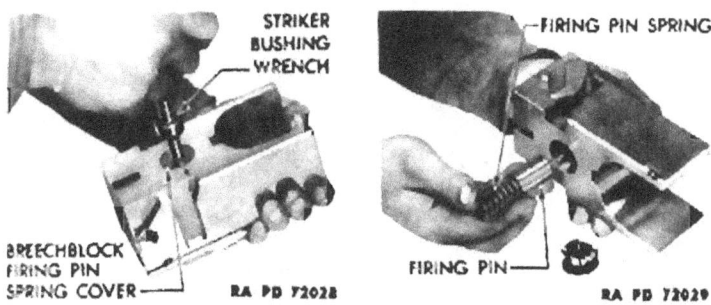

Figure 136 — Breechblock Firing Pin Spring Cover — Removal

Figure 137 — Breechblock Firing Pin and Spring — Removal

## 82. BREECHBLOCK DISASSEMBLY AND ASSEMBLY.

a. Firing Pin Removal. With the striker bushing wrench, remove the breechblock firing pin spring cover by pressing the wrench inwardly and turning it 90 degrees in either direction (fig. 136). Remove the breechblock firing pin spring and the firing pin (fig. 137).

b. Firing Pin Installation. Place the firing pin spring in the cup of the firing pin. Insert the spring and firing pin in the bore of the breechblock (fig. 137). With the striker bushing wrench in the opening in the back of breechblock firing pin spring cover, press the cover into the bore of the breechblock and turn the cover 90 degrees to match the outer point of the diamond on the cover with the arrow on the rear face of the breechblock (fig. 136).

c. Inner Cocking Lever and Check Plunger Removal.

(1) Press the check plunger inwardly and withdraw the outer cocking lever (fig. 138). Press the check plunger, turn the breechblock so that the front face is downward, and shake out the inner cocking lever (fig. 139). Withdraw the check plunger and check plunger spring (fig. 140).

d. Check Plunger and Inner Cocking Lever Installation. Insert the check plunger spring and check plunger in their bore in the breechblock (fig. 132). While holding the check plunger in, insert the inner cocking lever, fitting the end which is slotted into the slot in the check plunger. Press the check plunger in and insert the outer cocking lever, fitting the splined portion of the outer cocking lever shaft into the splineways in the hub of the inner cocking lever.

## 83. BREECH RING DISASSEMBLY AND ASSEMBLY.

a. Extractor Removal. Screw the extractor spindle and safety plunger key puller into the threaded hole in the end of the extractor spindle (fig. 141). Holding the extractors upright and the extractor

TM 9-252
83

**40-MM AUTOMATIC GUN M1 (AA) AND 40-MM ANTIAIRCRAFT GUN CARRIAGES M2 AND M2A1**

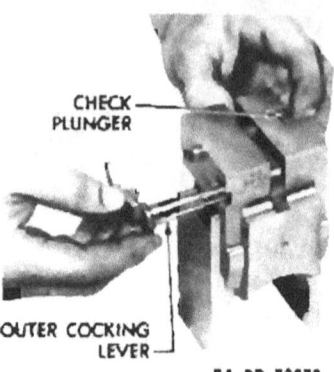

*Figure 138 — Outer Cocking Lever — Withdrawal*

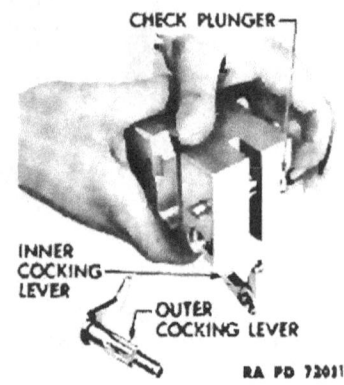

*Figure 139 — Inner Cocking Lever — Removal*

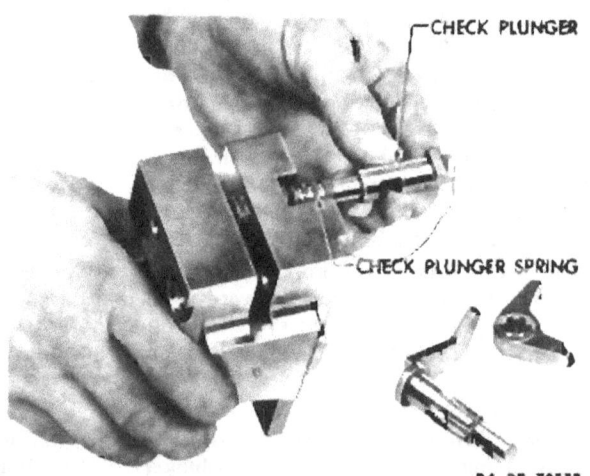

*Figure 140 — Check Plunger and Spring — Removal*

spindle arm down, withdraw the extractor spindle (fig. 142). Remove the extractor spindle arm and the extractors from the breech ring.

b. **Extractor Installation.** Place the extractor spindle arm in position. Assemble the two extractors together and lay them in position in the breech ring. Aline the keyways in the extractors and the key on the spindle arm by holding them in the manner shown in figure 142, and insert the extractor spindle.

174

## DISASSEMBLY AND ASSEMBLY

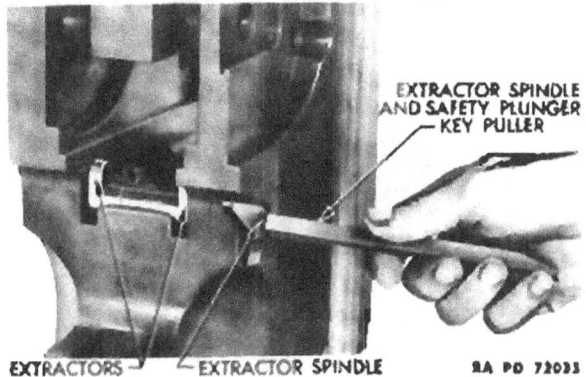

*Figure 141 — Puller Assembled to Extractor Spindle*

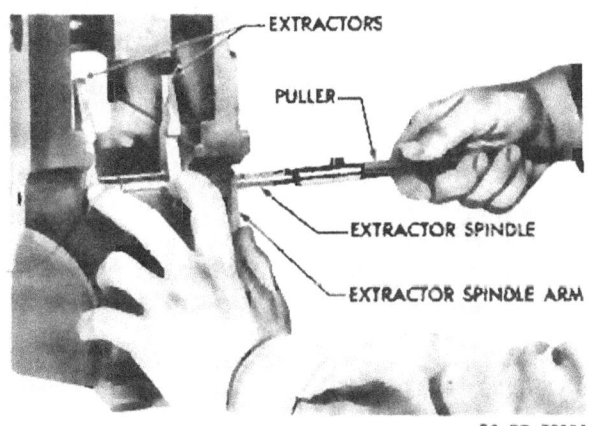

*Figure 142 — Extractor Spindle Installation*

NOTE: The extractors may be removed from the breech ring without removing the breech ring from the weapon. This operation is described in paragraph 80 b.

c. **Breech Ring Barrel Catch Removal.** Raise the breech ring barrel catch until the shaft which extends from the bushing is at right angles to the top surface of the breech ring. Withdraw the breech ring barrel catch control arm (fig. 143). In doing this, the key on the shaft of the arm must be mated with the keyway in the breech ring. Remove the barrel catch.

TM 9-252
83-84

**40-MM AUTOMATIC GUN M1 (AA) AND 40-MM ANTIAIRCRAFT
GUN CARRIAGES M2 AND M2A1**

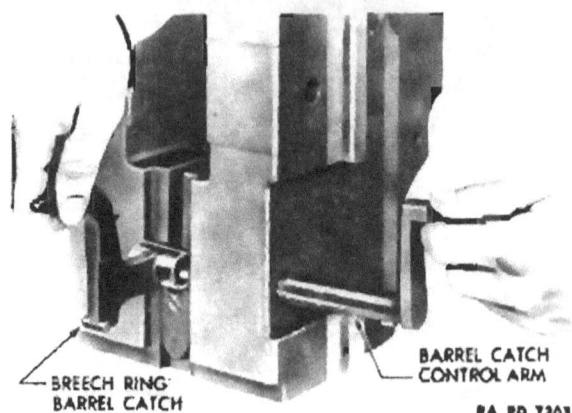

*Figure 143 — Breech Ring Barrel Catch — Removal*

*Figure 144 — Closing Spring Assembly on Bracket*

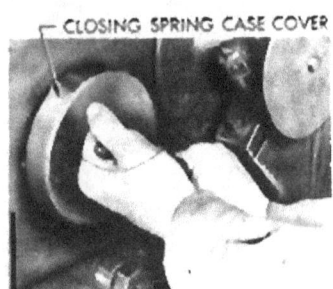

*Figure 145 — Closing Spring Cover — Release*

d. **Breech Ring Barrel Catch Installation.** Place the bushing of the barrel catch in the recess in the top of the breech ring, holding the barrel catch at right angles to the top surface of the breech ring. Insert the shaft of the barrel catch control arm into the breech ring through the bushing of the barrel catch.

**84. BREECH RING CLOSING SPRING DISASSEMBLY AND ASSEMBLY.**

a. **Disassembly.** Place the assembled breech ring closing spring case on the breech ring closing spring case bracket (fig. 144). This bracket is located on the right side of the breech casing near the rear end. Insert the closing spring cover wrench in the hole in the center of the case, engaging the protruding end of the closing spring

TM 9-252
84

## DISASSEMBLY AND ASSEMBLY

*Figure 146 — Closing Spring, Case, and Cover — Disassembled*

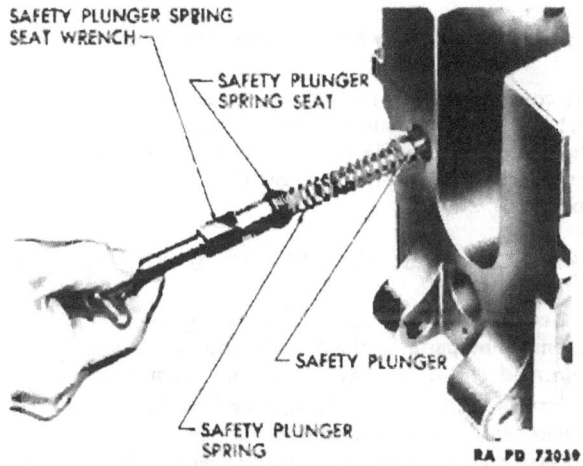

*Figure 147 — Breech Ring Safety Plunger — Disassembled*

with the pronged end of the wrench. Force the closing spring case cover inwardly to disengage the bayonet connection on the inside of the rim of the cover and on the outside of the rim of the case (fig. 145). Permit the cover to rotate counterclockwise under the action of the spring, taking care to prevent the tension from being released too violently. Remove the case cover and wrench (fig. 146). Remove the closing spring from the case.

h. **Assembly.** With the closing spring case on the closing spring case bracket, place the closing spring in the case (fig. 146). Fit the closing spring case cover over the case and spring. Insert the closing

177

TM 9-252
84-85

**40-MM AUTOMATIC GUN M1 (AA) AND 40-MM ANTIAIRCRAFT
GUN CARRIAGES M2 AND M2A1**

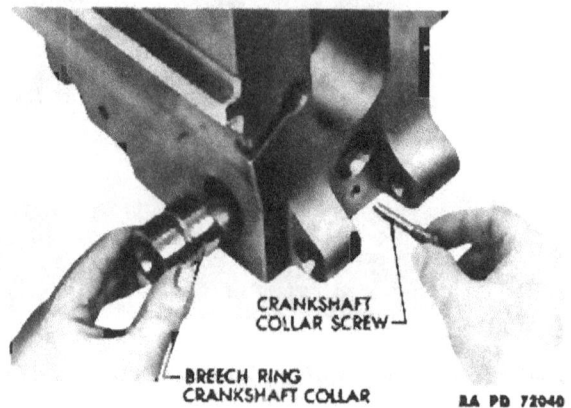

*Figure 148 — Breech Ring Crankshaft Collar — Disassembled*

spring cover wrench in the hole in the center of the case. Turn the wrench clockwise to wind the closing spring and to lock the bayonet connection of the case and cover. There will be an audible click as the case locks in position. Remove the closing spring assembly from the bracket.

## 85. BREECH RING BROKEN OR DEFORMED PARTS REPLACEMENT.

a. **General.** These instructions are given for the removal and replacement of parts of the breech ring which may become broken or deformed and of springs which may become broken or weak.

b. **Breech Ring Safety Plunger and Spring.** Unscrew the safety plunger spring seat by inserting the teats of the safety plunger and spring seat wrench in the slots of the seat and turning the seat counterclockwise. This seat is staked in position by having some of the metal of the breech ring driven into its slots. It must be worked out of its threaded bore carefully with short counterclockwise and clockwise twists to prevent damage to the seat or wrench. Remove the seat, safety plunger spring, and safety plunger (fig. 147). Replace defective parts with new ones, install, and stake.

NOTE: This operation should be performed only by qualified battery mechanic.

c. **Breech Ring Crankshaft Collar.** Unscrew and withdraw the breech ring crankshaft collar screw. Slide the breech ring crankshaft collar out of place (fig. 148). Replace worn or damaged collar with a new one and install parts.

## DISASSEMBLY AND ASSEMBLY

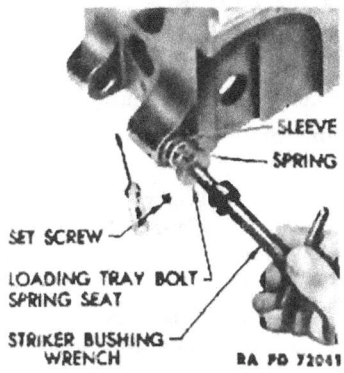

*Figure 149 — Loading Tray Bolt Spring — Disassembled*

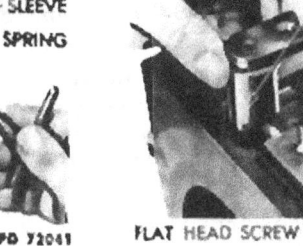

*Figure 150 — Breech Ring Barrel Abutment — Disassembled*

d. **Loading Tray Bolt Spring.** Remove the headless set screw from the edge of the loading tray bolt spring seat (fig. 149). Unscrew the seat, using the striker bushing wrench. Remove the loading tray bolt spring and sleeve. Replace damaged parts with new ones and install.

e. **Breech Ring Barrel Abutment.** Remove the flat-head screw which holds the breech ring barrel abutment in place in the top of the front end of the breech ring (fig. 150). Remove the abutment. Replace the damaged abutment with a new one and install.

## 86. FIELD STRIP ASSEMBLY.

a. **Breech Ring Installation.** Place the breech ring assembly in its guideways at the open rear end of the breech casing. With the top cover closed, slide the breech ring forward as far as it will go, fitting the lugs on the sides of the breech ring barrel catch into the slot on the under side of the top cover. Open the top cover and insert the breechblock locking pin into its holes in the breech casing and breech ring to keep the breech ring from sliding backward in the casing.

b. **Automatic Loader Installation.** With the loading tray installed in the loader (par. 81 g (8)), attach the front and rear loader lifters to the automatic loader. Lift the automatic loader and carefully place the front ends of the guides in the guideways of the breech casing. Slide the automatic loader forward until it is supported by the guideways. Remove lifters and slide the automatic loader forward until the loading tray attaching bolt can be inserted and locked with its bayonet connection.

c. **Rear Cover Attachment.** Raise the rear cover. Tap in the

## 40-MM AUTOMATIC GUN M1 (AA) AND 40-MM ANTIAIRCRAFT GUN CARRIAGES M2 AND M2A1

rear cover attaching bolt and replace the nut and cotter pin. Raise the lower end of the cartridge case deflector and insert the cartridge case deflector pin. Screw in automatic loader retaining screws.

d. **Barrel Assembly Installation.** Follow the instructions given in paragraph 79 d.

e. **Breech Mechanism Installation.** Install extractors (par. 80 c), if they have not already been installed. Elevate the gun to 45 degrees. Withdraw the breechblock locking pin. Install the breechblock, inner cranks, and breech ring closing spring case (par. 80 d).

f. **Recoil Cylinder Installation.** Holding the recoil cylinder at right angles to the bottom of the breech casing and with the flat surface on the recoil cylinder flange upward, engage the recoil cylinder piston rod retaining pin in the lugs on the lower front end of the breech ring. Raise the recoil cylinder and secure it in place with the recoil cylinder anchor bracket by screwing in the four anchor bracket cap screws. Rotate the recoil cylinder collar until it bears against the side of the bracket. Lock the collar with its set screw.

NOTE: Recoil cylinder anchor bracket must be correctly mounted, or damage to recoil mechanism will result. Anchor brackets must be cleaned so numbers on casting will show. In installing the anchor bracket, the numbers on the anchor bracket must be matched with those on the breech casing.

## 87. EQUILIBRATOR DISASSEMBLY AND ASSEMBLY.

a. **Disassembly.**

(1) Open the equilibrator case cover and insert the springholding U-bar (fig. 151).

(2) Remove the equilibrator rod jam nut and bushing nut with equilibrator rod bushing nut wrench (fig. 152). At this point, the equilibrator springs will have expanded to the extent that the equilibrator spring retainer will be forced against the U-bar.

(3) Attach the equilibrator spring compressor to the end of the equilibrator spring rod (fig. 153). Screw the equilibrator spring compressor sufficiently to force the equilibrator spring retainer away from the U-bar so that the U-bar may be removed. Remove the U-bar.

(4) Unscrew the spring compressor, relieving the compression on the equilibrator springs (fig. 154). When the compression has been entirely relieved, disassemble the spring compressor from the equilibrator rod. Remove the retainer, springs, and separators from the equilibrator case.

b. **Assembly.**

(1) Remove the handle from the equilibrator spring compressor.

TM 9-252
87

## DISASSEMBLY AND ASSEMBLY

Figure 151 — Spring-holding U-bar — Insertion

Figure 152 — Equilibrator Rod Bushing Nut — Removal

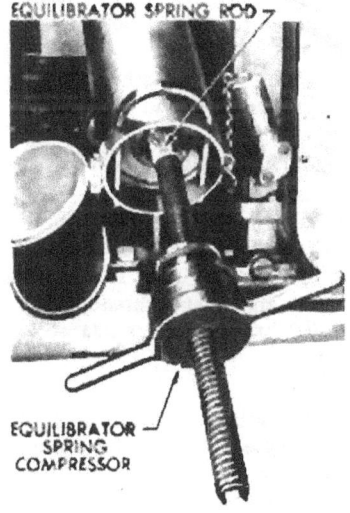

Figure 153 — Equilibrator Spring Compressor — In Place

Figure 154 — Equilibrator Spring Rod — Removal

TM 9-252
87-88

**40-MM AUTOMATIC GUN M1 (AA) AND 40-MM ANTIAIRCRAFT GUN CARRIAGES M2 AND M2A1**

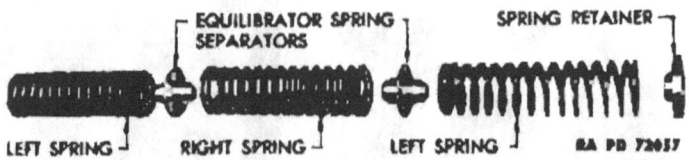

*Figure 155 — Equilibrator Springs, Separators, and Retainer*

Attach the threaded shaft of the compressor to the end of the equilibrator spring rod.

(2) Place the equilibrator springs, separators, and retainer on the threaded shaft of the compressor and the equilibrator spring rod in the following order: one left spring; one separator; one right spring; one separator; one left spring; one retainer (fig. 155).

(3) Screw on the compressor handle and compress the springs sufficiently to permit the U-bar to be installed. Remove the compressor. Screw on the equilibrator rod bushing nut. Remove the U-bar.

(4) After both equilibrators have been assembled to this point, test the amount of effort required to elevate and depress the gun. If the effort to depress the gun is more than that required to elevate, turn out the bushing nut; if the effort to elevate is more than that to depress the gun, turn in the bushing nut. Adjust both assemblies exactly alike.

(5) When the adjustment is completed, lock both bushing nuts in position with the equilibrator rod jam nuts. Close end covers.

## 88. WHEEL, HUB, AND BRAKE DRUM DISASSEMBLY AND ASSEMBLY.

a. **Wheel Removal.** Lift the wheels clear of the ground by raising the carriage with the leveling jacks. Remove the wheel retaining nuts with the wheel stud nut wrench (fig. 156). Wheels on the right side of the carriage have right-hand thread studs and nuts; wheels on the left side have left-hand thread studs and nuts. Remove the wheel and tire assembly.

b. **Hub and Brake Drum Removal.**

(1) Remove the hub cap with the hub cap wrench (fig. 157).

(2) Remove the cotter pin from the wheel retaining castle nut on the axle spindle (fig. 158). Unscrew the castle nut. Remove the axle spindle washer and outer roller bearing. Dismount the brake drum and hub assembly from the axle spindle. Remove the inner roller bearing (fig. 159).

CAUTION: Do not allow grease or oil to touch the magnet or

## DISASSEMBLY AND ASSEMBLY

*Figure 156 — Wheels — Removal*

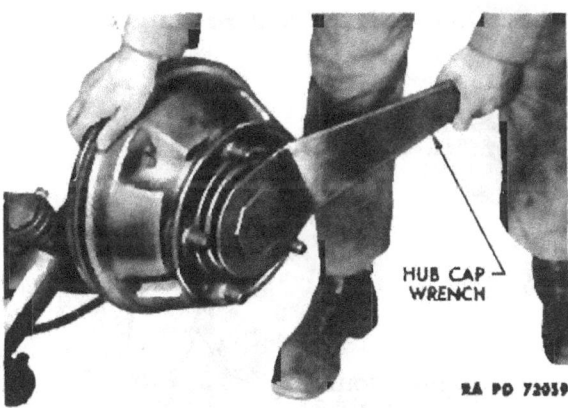

*Figure 157 — Hub Cap — Removal*

brake linings. Do not touch the magnet or brake linings with oily hands.

c. **Hub and Brake Drum Installation.**

(1) Clean drum, armature, wheel bearings, nut, washer, hub cap, hub, and axle spindle of old grease. Wash with SOLVENT, dry-cleaning.

CAUTION: Do not allow any grease or solvent to touch the brake

**TM 9-252**
**88**

**40-MM AUTOMATIC GUN M1 (AA) AND 40-MM ANTIAIRCRAFT GUN CARRIAGES M2 AND M2A1**

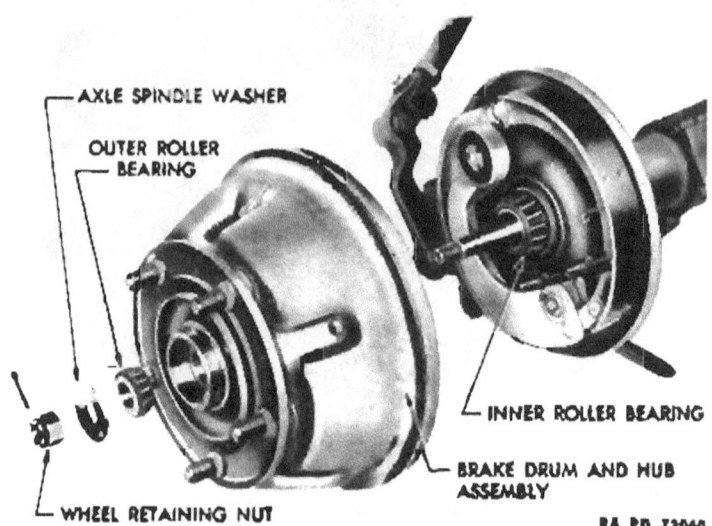

*Figure 158 — Wheel Hub Parts — Disassembled View*

*Figure 159 — Inner Roller Bearing — Removal*

## DISASSEMBLY AND ASSEMBLY

lining or magnet. The brake drum and armature must be clean and free of grease.

(2) If the brake magnet and armature show signs of wear, or if the brake lining or magnet are glazed, the matter should be brought to the attention of the ordnance maintenance personnel for correction.

(3) Be sure that the solvent is dried off the wheel bearings, then repack the bearings with grease in the prescribed manner.

(4) Install the grease-packed inner wheel bearing on the spindle. Install the hub and drum assembly. Install the grease-packed outer wheel bearing in the hub. Slide the axle spindle washer on the spindle.

d. **Wheel Bearing Adjustment.** Screw on the wheel retaining castle nut. Tighten the nut and rotate the brake drum. Tighten until a drag is felt when the hub and drum assembly is rotated. Then back off the nut until the hub and drum assembly rotate freely; one-sixth turn is usually enough. Fit the cotter pin in the wheel retaining castle nut. Install the hub cap.

e. **Wheel Installation.** Install the wheel and wheel retaining nuts. Tighten the nuts in rotation to center the studs in their holes in the wheel, using the wheel stud nut wrench. Tighten the nuts securely.

## 89. TIRE REMOVAL AND INSTALLATION.

NOTE: For maintenance and care of pneumatic tires and rubber treads, refer to TM 31-200.

a. **Divided Rim Type Wheels.**

(1) REMOVAL.

(a) Remove the wheel from the carriage (par. 88 a). Remove the valve cap and valve core.

CAUTION: Tires must be completely deflated before any attempt is made to remove wheel rim flange retaining nuts. An inflated tire may blow a partially removed flange off the wheel and cause serious injury to personnel.

(b) Remove all of the flange retaining nuts. Drive a tire iron between the flange and tire; then lift off the flange.

(c) Turn over the wheel and tire, loosen the wheel from the tire by prying with a tire iron, and lift the wheel out of the tire.

(d) To remove the beadlock from combat tires, inflate the tube until the tire beads spread away from the flanged blocks. If the beadlock flange blocks stick to the tire, pry them loose with a tire iron. Pull out the beadlock.

(e) Deflate the tube. Remove the flap and tube.

(2) INSTALLATION.

(a) Be sure that all dirt and rust have been removed from the tire and that the tube is clean; then insert the tube and flap. The

### 40-MM AUTOMATIC GUN M1 (AA) AND 40-MM ANTIAIRCRAFT GUN CARRIAGES M2 AND M2A1

flap must be smooth; folds or wrinkles will result in tube failure.

*(b)* On combat tires, inflate the tube to spread the tire beads to install the beadlock. Place the beadlock over the valve. Push the beadlock down between the beads. Fold the flexible band of the beadlock and insert it between the tire beads. Deflate the tube, making sure that flanged blocks of the beadlock are centered between the beads.

*(c)* Place the tire assembly on the wheel, centering the valve in the valve slot. Install the wheel rim flange and tighten four nuts equally spaced around the wheel; this is to seat the flange evenly. Then install the remaining nuts and tighten all nuts.

*(d)* Install the valve core and inflate the tire to 45 pounds per square inch. Screw on the valve cap tightly by hand.

b. **Flat Base Rim Type Wheels.**

(1) REMOVAL.

*(a)* Remove the wheel from the carriage (par. 88 a). Remove the valve cap and valve core.

CAUTION: Make sure the tire is completely deflated.

*(b)* Using two tire irons, pry the two ends of the locking ring apart, and at the same time, pry the slotted end of the ring out of the rim well. With the slotted end of the ring free of the rim, continue to pry or pull the rest of the ring out of the rim well, being careful not to bend or spring the ring out of round. (To remove the beadlock from combat tires, see par. 89 a (1) *(d)*.)

*(c)* Turn the wheel and tire assembly over and block up under the wheel. Pry the tire bead loose from the wheel, using a tire iron and hammer if necessary. Force the tire down off the wheel. Lift off the wheel assembly.

*(d)* Remove the tire inner liner or flap. Spread the tire beads apart and work the tube out of the tire casing.

(2) INSTALLATION.

*(a)* Be sure that all dirt and rust have been removed from the tire and that the tube is clean. A small amount of tire talc applied to the inside of the tire is recommended. Insert the tube and flap, working the edges of the flap under the tire beads. The flap must be smooth; folds or wrinkles will result in tube failure. (To install tubes in compact tires, see par. a (2) *(b)*.)

*(b)* Insert the plain end of the locking ring in the rim well. Pry down on the ring about one-third of the distance around the ring from the plain end. At the same time, hammer more of the ring into the rim well. Spread the locking ring by prying the two ends apart, and while doing this, hammer the ring into the rim well throughout its entire length. Make sure the ring is securely seated in the rim well before inflating the tire.

## SIGHTING AND FIRE CONTROL EQUIPMENT

(c) Install the valve core and inflate the tire to the recommended pressure, 45 pounds per square inch. Install the wheel (par. 88 e).

CAUTION: Tap the locking ring during the initial inflation of the tire to seat it firmly in the rim well. Stand aside while inflating the tire to avoid personal injury in case the locking ring is not properly seated in the rim well and flies off the wheel.

### Section X

## SIGHTING AND FIRE CONTROL EQUIPMENT

**90. ARRANGEMENT OF SIGHTING AND FIRE CONTROL EQUIPMENT.**

a. This section describes the on-carriage sighting and fire control equipment. It describes the operation, test, maintenance, and care and preservation of this equipment. Off-carriage parts, including Generating Unit M5, the Director M5A2, M5A1, or M5, the Cable System M8, Gunner's Quadrant M1 or M1918, and cable repair kit, are covered in other Technical Manuals (sec. XV).

NOTE: The arrangement of the azimuth and elevation mechanisms of this weapon is opposite from the usual arrangement of these mechanisms on other artillery weapons in that the azimuth mechanism is on the right side and the elevation mechanism is on the left side.

b. The fire control equipment is designed to operate as a coordinated system under the conditions encountered in short range antiaircraft fire on fast moving targets. In the event that the fire control equipment, or any part of it, is unavailable or disabled, emergency sighting equipment is provided.

c. The sighting equipment includes the Computing Sight M7 or M7A1, Sighting System M3, and bore sights.

(1) The elevation and azimuth operators sight through the rear and front open sights of the Sighting System M3 to track the target and to set in the necessary lead.

(2) The elevation and azimuth trackers follow the target through telescopes of the Computing Sight M7 or M7A1. Vertical and lateral deflections are mechanically set into the telescopes by a mechanism which is subject to adjustment by a sight setter who stands behind the azimuth tracker.

(3) Bore sights and a gunner's quadrant are provided for use in performing bore sighting adjustments and in orienting.

d. Gun pointing data originate at the AA Director M5A2, M5A1, or M5. The director determines the target position in space from the angular position of two tracking telescopes with which the opera-

**TM 9-252**

**90**

**40-MM AUTOMATIC GUN M1 (AA) AND 40-MM ANTIAIRCRAFT GUN CARRIAGES M2 AND M2A1**

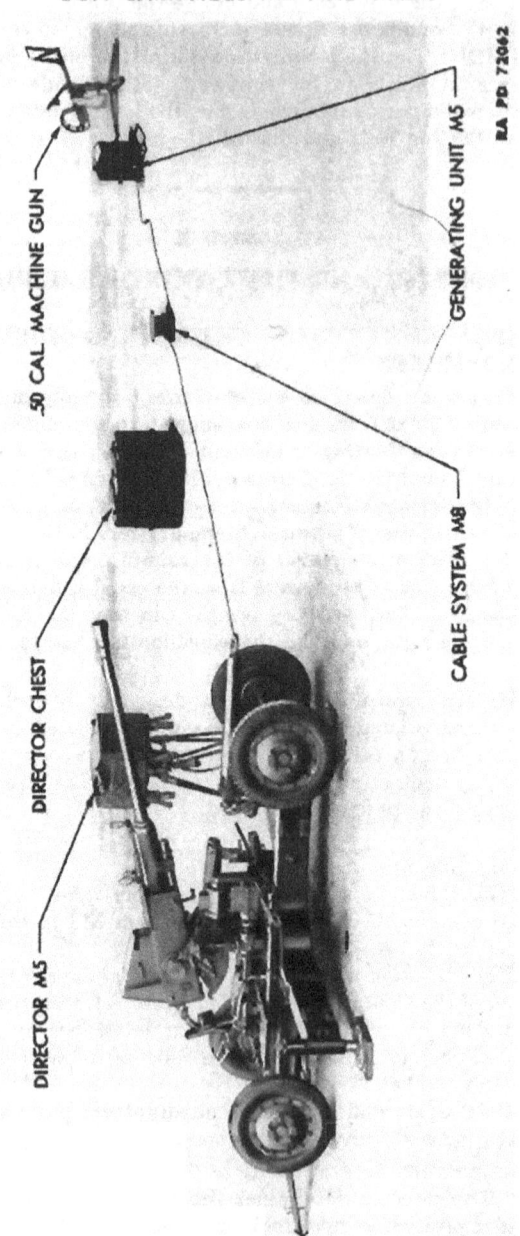

Figure 160 — Sighting and Fire Control Equipment

## SIGHTING AND FIRE CONTROL EQUIPMENT

tors follow (track) their target. One telescope gives the horizontal position or azimuth; the other, the vertical angle or elevation. Ranges are estimated by the range setter at the director and may be adjusted quickly from observation of the tracer stream during fire. Smooth and accurate tracking at high angular rates of travel is made possible by use of a system of tracking known as aided tracking. The director produces firing azimuth and quadrant elevation data for transmission to the gun.

NOTE: The Director M5A2 will supersede Directors M5 and M5A1.

e. The Remote Control System M5 points the gun in azimuth and elevation according to the controlling data furnished from the director. The system includes electrical and hydraulic power equipment (oil gears) mounted on the carriage and connected to the traversing and elevating mechanisms.

f. Electric power to operate the director and power mechanisms for remote control is supplied by a gasoline-electric a-c Generating Unit M5.

## 91. DIRECT FIRE SIGHTS.

a. Guns in service are equipped with two types of direct fire sights. Guns not equipped with the Computing Sight M7 or M7A1 are equipped with Sighting System M3 (fig. 161). Guns equipped with the Computing Sight M7 are also provided with direct fire sights for emergency use and for bore sighting operations (fig. 163).

b. Sighting System M3, Guns Not Equipped With Computing Sight M7.

(1) These sights (fig. 161) are used when the director is not available. They provide a simple means of giving the necessary lead to allow for target travel during the time of flight of the shell, and of changing the lead quickly during the engagement owing to the change of position of the target.

(2) A "wheel" type front sight is used for elevation, and a "gate" type front sight is used for azimuth. The wheel type of sight has a clock face on which the course of the target is interpreted in terms of a clock hour. The gate type has vertical bars which are spaced in terms of degrees.

(3) The rear elevation sight has an adjusting lever for raising and lowering the line of sight.

(4) The front elevation sight and the front azimuth sight are mounted on slides which permit lateral adjustment. Each slide is secured by two wing nuts.

(5) When using these sights, the traversing and elevating hand cranks must be engaged (pushed in) and the oil gears of the remote control system must be disengaged. To disengage the azimuth oil

TM 9-252
91

**40-MM AUTOMATIC GUN M1 (AA) AND 40-MM ANTIAIRCRAFT GUN CARRIAGES M2 AND M2A1**

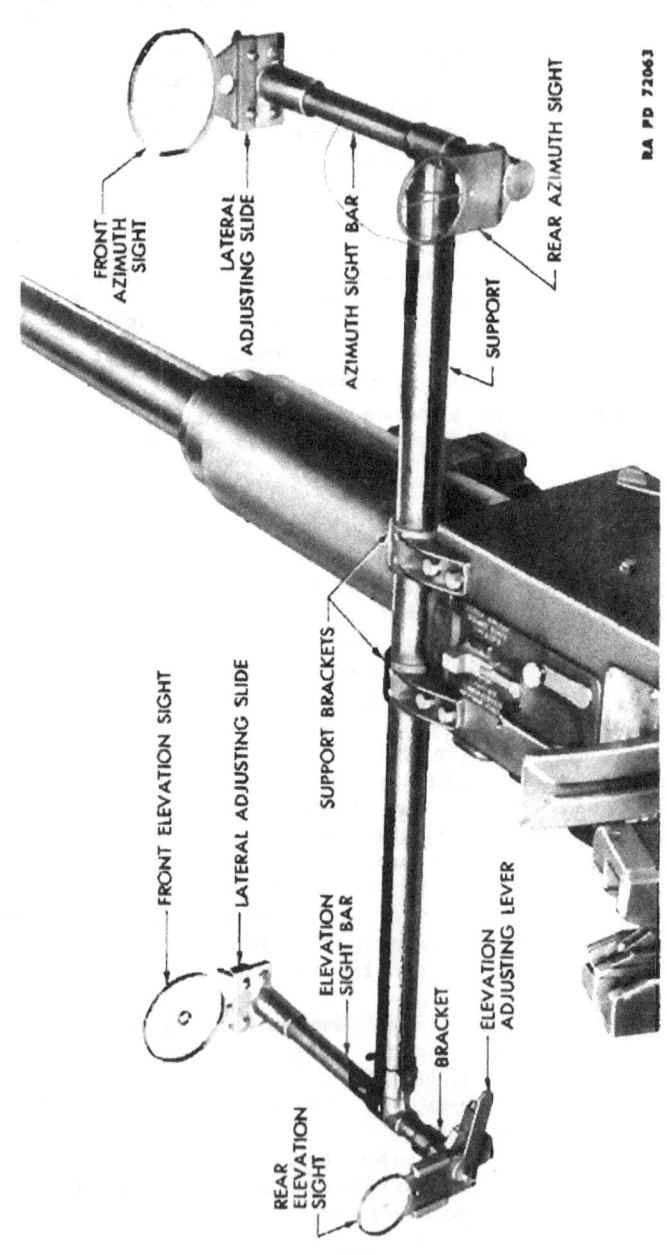

Figure 161 — Sighting System M3 — Gun Not Equipped With Computing Sight M7 or M7A1

## SIGHTING AND FIRE CONTROL EQUIPMENT

gear, disengage the power synchronizing mechanism by pulling the power synchronizing handle upward. This shuts off power to the oil gear and disengages the azimuth oil gear clutch. To disengage the elevation oil gear, move the elevation limit switch to the "OFF" position, disconnecting power to the elevation oil gear. Then disengage the elevation oil gear clutch lever by lifting the clutch lever lock and moving the clutch lever to the disengaged position (the top of the clutch lever toward the output coupling).

(6) Leads are set by tracking the target at different points on the front sight cross wires, in accordance with the practice of the using arms.

(7) The sights are ruggedly constructed, but because of their exposed position, must be protected against damage by bending. The sights should not be used as handholds or supports. Keep wing nuts securely tightened. Keep the cam surface under the elevation adjusting lever clean.

c. **Direct Fire Sights, Guns Equipped With Computing Sight M7 or M7A1.**

(1) These sights (fig. 163) are for use when neither the director nor Computing Sight M7 or M7A1 is available or in operating condition. They provide a simple means of estimating the necessary lead to allow for target travel during the time of flight of the shell.

(2) The rear sights (B, fig. 163) on both the azimuth and elevation sides of the gun are subject to adjustment for bore sighting. Vertical adjustment is accomplished by raising or lowering the peep rings in their holders. Elongated holes in the base of the holders provide means of making lateral adjustments.

(3) The front sights (A, fig. 163) on both the azimuth and elevation sides of the gun are of the "speed ring" type; permitting the course of the target to be interpreted in terms of a clock hour.

(4) When in use, the front sights are secured in their slides by two wing nuts.

(5) When using these sights, the traversing and elevating hand cranks must be engaged (pushed in) and the oil gears of the remote control system must be disengaged. To disengage the azimuth oil gear, disengage the power synchronizing mechanism by pulling the power synchronizing handle upward. This shuts off power to the oil gear and disengages the azimuth oil gear clutch. To disengage the elevation oil gear, move the elevation limit switch to the "OFF" position, disconnecting power to the elevation oil gear. Then disengage the elevation oil gear clutch lever by lifting the clutch lever lock and moving the clutch lever to the disengaged position (the top of the clutch lever toward the output coupling).

(6) Leads are set by tracking the target at different points on the front sight cross wires, in accordance with the practice of the using arms. The outer ring is equivalent to a target speed (in a plane at

**TM 9-252**
**91-92**

**40-MM AUTOMATIC GUN M1 (AA) AND 40-MM ANTIAIRCRAFT GUN CARRIAGES M2 AND M2A1**

right angles to the axis of the bore of the gun) of 300 miles per hour; the middle ring, 200 miles an hour; the inner ring, 100 miles per hour; and the small vertical wires on the horitizontal ("9 o'clock-3 o'clock") cross wire, 20 miles per hour. The vertical distance between the "9 o'clock-3 o'clock" wire and the small horizontal wire on "12 o'clock" is equivalent to superelevation for normal range. For very close ranges, therefore, the target should be sighted on this small wire, thereby removing superelevation. This wire is also used when bore sighting these sights.

(7) The front sights should *not* be mounted when Computing Sight M7 or M7A1 is in operating condition (fig. 163). When mounting them, be sure that the slides and stop surfaces are clean. Push the front sights down firmly against their stops and then securely tighten the wing nuts.

## 92. COMPUTING SIGHTS M7 AND M7A1.

**a. General.**

(1) This sighting system (figs. 162, 163, and 164) provides automatic correction of lateral and vertical deflection of its telescopes (M7 on the azimuth side and M74 on the elevation side) as the gun is moved in azimuth and elevation. Initial settings for course and speed of target, and subsequent adjustments thereof, are made by hand. The M7 and M7A1 Sights have been designed primarily for antiaircraft use but are equally effective in establishing leads on relatively slower moving ground targets.

(2) The Computing Sight M7A1 is exactly the same as the Computing Sight M7, except that the M7 model has a larger azimuth gear box handwheel. In addition, a different method of bore sighting and adjusting of the vertical axis of the deflection mechanism control case is necessary, due to a slight change in the link connecting the superelevation cam shaft arm to the trunnion frame.

**b. Description.**

(1) The telescope supporting assembly is bolted to the upper surface of the forward end of the breech casing. On the azimuth side of this assembly are the mechanisms which provide proper vertical and lateral deflections, including superelevation. Deflections given the azimuth telescope (fig. 162) are imparted to the elevation telescope (D) by a system of links and torque rods, the latter passing through the main support bar (E). Orientation of the direction of lead indicating arrow (H) is maintained as the gun is turned in azimuth by a drive through a flexible shaft (F) from the azimuth handwheel gearbox to the deflection mechanism of the main assembly. The link (O) from the superelevation cam shaft arm to the stud on the gun trunnion frame maintains the deflection mechanism control case (J) level at all elevations of the gun, and automatically

# TM 9-252
## SIGHTING AND FIRE CONTROL EQUIPMENT

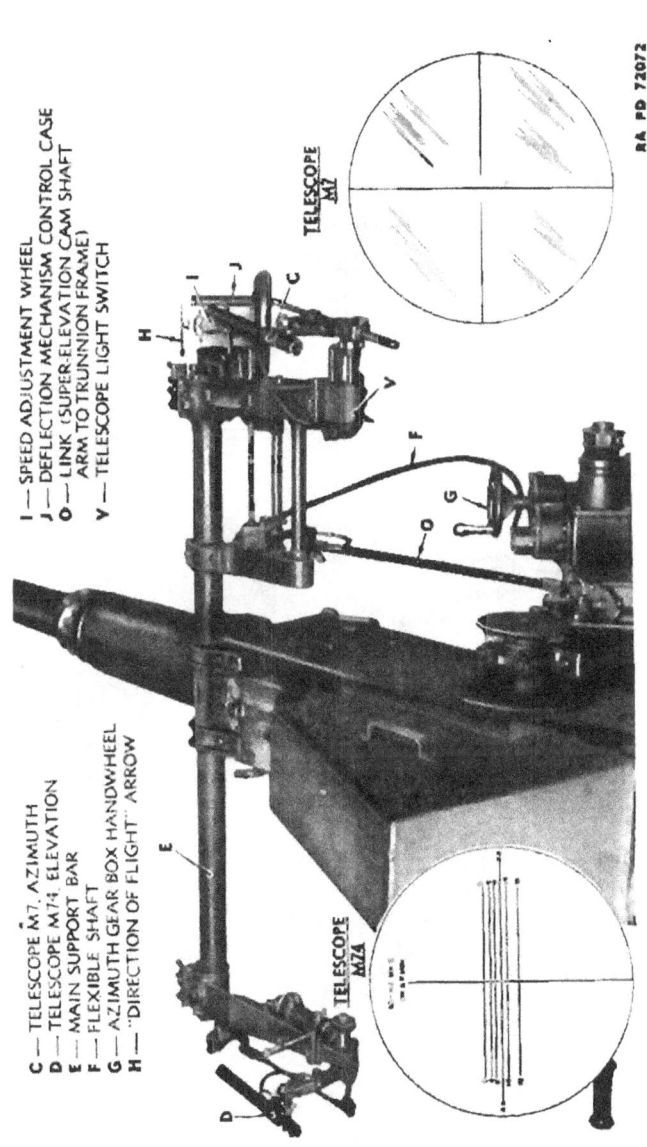

C — TELESCOPE M7, AZIMUTH
D — TELESCOPE M74, ELEVATION
E — MAIN SUPPORT BAR
F — FLEXIBLE SHAFT
G — AZIMUTH GEAR BOX HANDWHEEL
H — "DIRECTION OF FLIGHT" ARROW
I — SPEED ADJUSTMENT WHEEL
J — DEFLECTION MECHANISM CONTROL CASE
O — LINK (SUPER-ELEVATION CAM SHAFT ARM TO TRUNNION FRAME)
V — TELESCOPE LIGHT SWITCH

Figure 162 — Computing Sight M7 Ready for Use With Telescopes M7 (Azimuth — Right) and M74 (Elevation — Left) in Their Holders

**TM 9-252**
**92**

**40-MM AUTOMATIC GUN M1 (AA) AND 40-MM ANTIAIRCRAFT GUN CARRIAGES M2 AND M2A1**

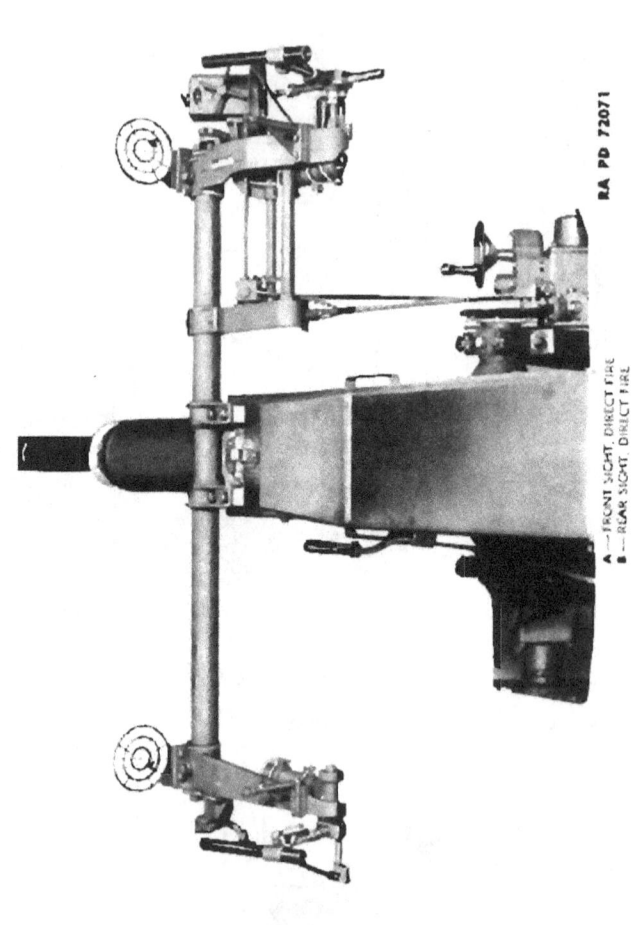

Figure 163 — Direct Fire Sights Mounted on Computing Sight M7 Main Assembly

## SIGHTING AND FIRE CONTROL EQUIPMENT

causes adjustment of superelevation and lead as the gun moves in elevation.

(2) Direction of lead is indicated by the arrow mounted on the top of the deflection mechanism control case. The direction of the arrow, and thus of the lead, may be adjusted by turning the arrow itself or by turning the handwheel (G) on the azimuth gearbox to which the lower end of the flexible shaft is attached. The arrow and handwheel rotate in the same direction relative to the gun, but opposite to the direction of the traverse, thereby maintaining orientation of the arrow with respect to the target path.

(3) Adjustment for speed of the target is made by turning the spoked wheel (I) mounted on the face of the deflection mechanism control case. The figures on the dial show the level flight speed set into the mechanism.

(4) Effects of change in the speed setting and in the direction of the arrow with respect to the gun are conveyed to the telescope deflection system by the telescoping shaft attached to the swivel extending from the lead screw housing (K) on the undersurface of the deflection mechanism control case.

(5) Computing Sights M7 and M7A1 mount two Tracking Telescopes, M7 on the azimuth side and M74 on the elevation side, which are identical except for difference of reticle.

(a) Both are erecting telescopes with the following optical characteristics:

| | |
|---|---|
| Power | 1 x |
| Field of view | 11 deg |
| Diameter of exit pupil | 0.6 in. |
| Eye distance | 4.384 in. |

(b) Each telescope is clamped in a holder from which it is readily removed for traveling. Each holder has a locating projection to insure correct vertical and horizontal positioning of the reticle lines.

(c) The Telescope M7 (C, in fig. 162) is so mounted on the azimuth side as to show a solid vertical line and a broken horizontal line of the reticle. The azimuth tracker keeps the vertical line on the target.

(d) The Telescope M74 (D, in fig. 162) is so mounted on the elevation side as to show several horizontal lines graduated to give the proper superelevation of the gun when used against land targets at various ranges. For antiaircraft, the elevation tracker keeps the line marked "A-A" on the target.

(e) A dust cap, which consists of a leather cap for each end, connected by a strap, is provided for each telescope. One packing chest, furnished with padlock and keys, with compartments for both telescopes, is provided for each carriage.

(f) The lighting device provided for each telescope includes a metal tube containing two standard flashlight cells, a switch, and

TM 9-252
92

**40-MM AUTOMATIC GUN M1 (AA) AND 40-MM ANTIAIRCRAFT GUN CARRIAGES M2 AND M2A1**

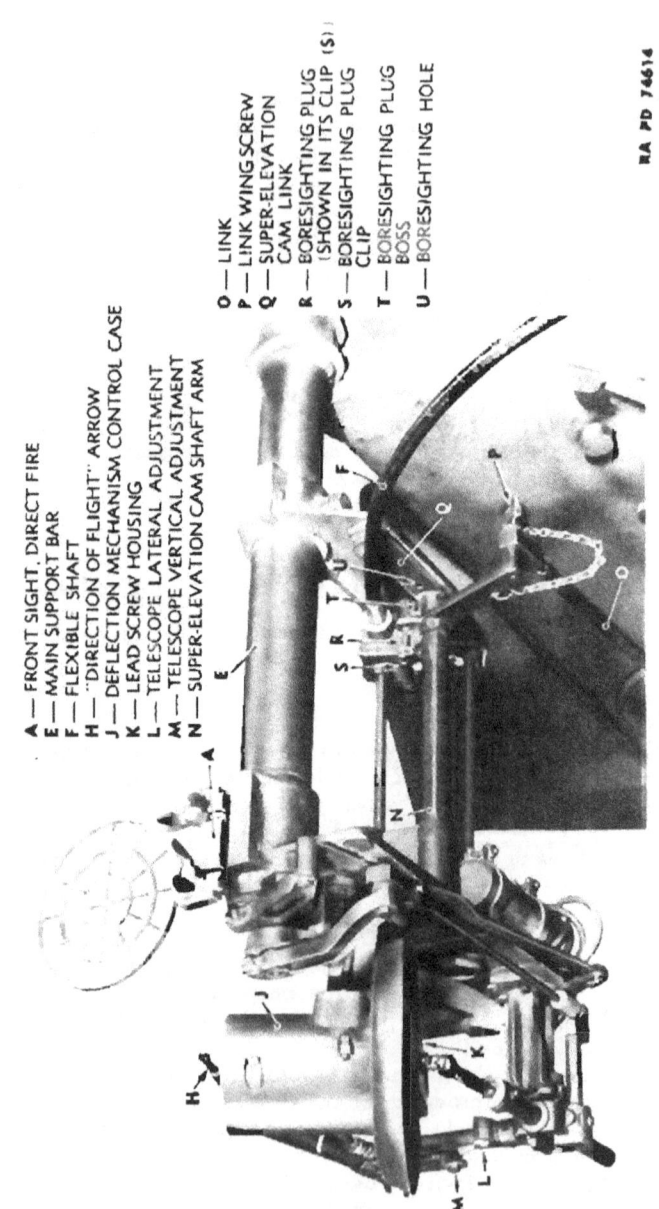

Figure 164 — Azimuth Side of Computing Sight M7 With Direct Fire Sight Mounted

196

## SIGHTING AND FIRE CONTROL EQUIPMENT

a flexible cord with plug to fit in the lamp socket of the telescope holder. The metal tube fits into a sleeve on the arm of the sighting system.

c. Operation.

(1) Clamp the Telescopes M7 and M74 in their holders. In each case, the locating slot of the telescope is positioned so that the broken line is the line not required for tracking.

(2) When necessary, illuminate each reticle by closing the switch in the front end of the lighting device.

(3) Engage (push in) the traversing and elevating hand cranks.

(4) Disengage the oil gears by pulling up the azimuth oil gear clutch operating handle, throwing the elevation limit switch to "OFF" position, and disengaging the elevation oil gear clutch lever (move top of lever toward coupling), locking it by means of the auxiliary lever.

(5) As the azimuth and elevation trackers slew the gun toward a target, a sight setter (standing behind the azimuth tracker) turns the speed wheel (I) to what he estimates to be the speed of the target (which registers on the dial) and turns the arrow about its vertical axis so that it lies in the slant plane containing the gun and the target path. (If the target path is level, the arrow shaft should be parallel with it. If the target is diving, the head of the arrow should point to the spot where the target path, if extended, would strike the ground (or horizontal plane). If the target is climbing, the head of the arrow should be pointed directly away from the spot where a rearward extension of the target path would intersect the horizontal ground plane.)

(6) The trackers aline their telescopes on the target and so track as to continue to maintain their respective cross hairs thereon.

(7) The sight setter observes the tracer stream and makes such adjustments of the arrow and speed setting as to pass the trajectory curve through the target.

d. Tracking Adjustments.

(1) ADJUSTING SPEED.

(a) If the arrow is positioned with respect to the target path (subpar. c (5), above), all shells will pass through the path of the target regardless of the speed setting. However, this "tracer cross" will not occur at the target (be a hit) unless the proper speed has been set into the sight. Note whether the tracer stream crosses the line of sight to the target between the gun and the target, or passes beyond the target and is momentarily blotted out by the target (fig. 165).

(b) If the tracer stream passes on the near side of the target; increase the speed setting.

(c) If the tracer stream passes on the far side of the target; reduce the speed setting.

**40-MM AUTOMATIC GUN M1 (AA) AND 40-MM ANTIAIRCRAFT GUN CARRIAGES M2 AND M2A1**

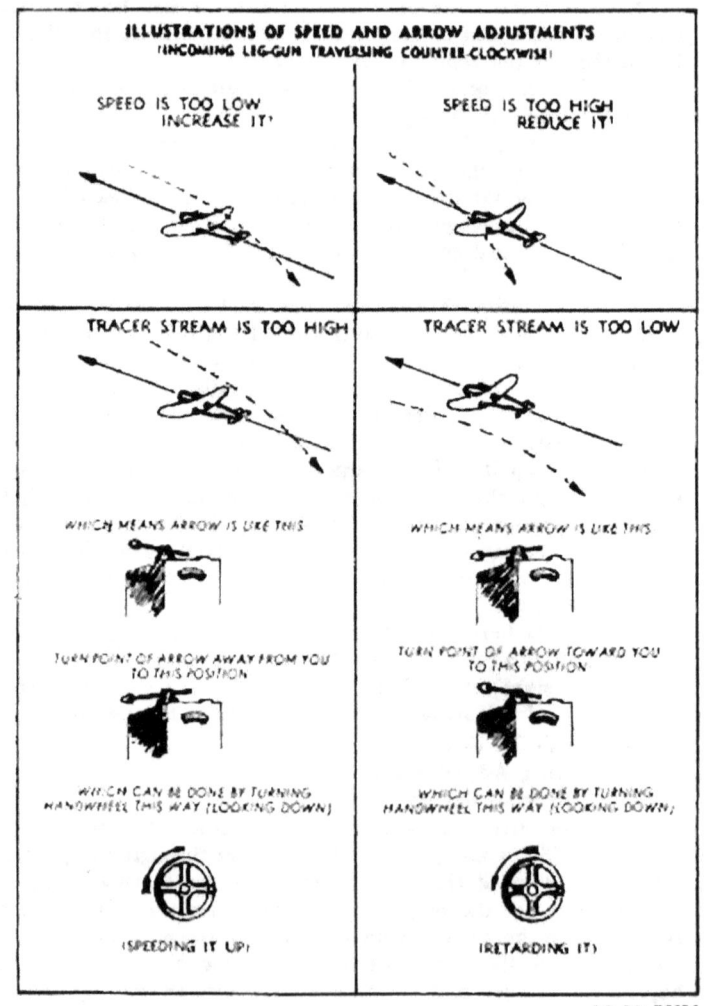

Figure 165 — Adjustment of Fire by Observation of the Tracer Stress

## SIGHTING AND FIRE CONTROL EQUIPMENT

(2) ADJUSTING SPEED FOR OTHER THAN HORIZONTAL FLIGHT. For other than horizontal flight, speed must be constantly changed during the course. In the case of a diving target, start with a lower than estimated speed for the incoming leg; increase to what is believed the correct speed just before the midpoint of the course, and then to an increasingly greater speed for the outgoing leg. In the case of a climbing target, start the incoming leg with a greater than estimated speed and then constantly decrease the speed setting, as indicated by the tracer stream.

(3) ADJUSTING DIRECTION OF LEAD INDICATING ARROW.

*(a)* If the course is subject to cross wind or the arrow is not correctly set, the tracer stream will not cross the line of sight to the target, but will seem either high or low with respect to the target as it passes the line of sight to the target (fig. 165).

*(b)* If the tracer is high with respect to the target, rotate the arrow so that its head moves away from the sight setter slightly. This can be done either by grasping and turning the arrow itself, or by rotating the handwheel in the opposite direction to the turn of the gun in azimuth (by speeding up the rotation of the handwheel).

*(c)* If the tracer is low with respect to the target, raise it by rotating the arrow so that its head moves toward the sight setter. This can be done by either grasping and turning the arrow itself, or by rotating the handwheel in the same direction as the turn of the gun in azimuth (by retarding the rotation of the handwheel).

*(d)* On directly incoming or outgoing courses, the tracer stream may be shifted to the left by moving the head of the arrow to the left, and to the right by moving the head of the arrow to the right.

*(e)* Note that if the arrow is observed from below and to the rear as it is adjusted, the head of the arrow seems to rise slightly as it is rotated toward the sight setter. Thus, the tracer stream is shifted in the same direction as the head of the arrow for all conditions outlined in substeps *(b)*, *(c)*, and *(d)*, above.

*e.* **Disassembly and Assembly.** Except for the installation and removal of the telescopes or the forward emergency direct fire sights incident to normal use, the only other assembly or disassembly operations performed by the using arms are those necessary to renew the lamps or dry cells of the lighting device.

(1) To replace the dry cells, grasp the cap at the forward end of the lighting device and pull it forward, removing the tube containing the dry cells from the cylindrical sleeve on the arm. Unscrew the tube and replace the dry cells, placing both cells with their central terminals toward the cap. Use two standard flashlight battery cells. Screw the tube into the cap and replace it in the sleeve on the arm.

(2) To replace a lamp, remove the plug on the flexible cord from the lamp socket; then unscrew the socket, rendering the lamp acces-

TM 9-252
92

**40-MM AUTOMATIC GUN M1 (AA) AND 40-MM ANTIAIRCRAFT GUN CARRIAGES M2 AND M2A1**

Figure 166 — Azimuth Side of Computing Sight M7 in Bore Sighting Position

## SIGHTING AND FIRE CONTROL EQUIPMENT

sible. Replace the lamp, using a standard screwbase flashlight lamp, and return the socket and plug to their original positions.

### 93. TESTS AND ADJUSTMENTS OF COMPUTING SIGHTS M7 AND M7A1.

*a.* General. The alinement of the sighting system should be verified frequently and adjusted if necessary.

*b.* Alinement of Telescopes. The optical axis of the telescopes are alined parallel with each other and with the bore of the gun when superelevation and all deflections are removed from the sight as explained in step (1), below, for the Computing Sight M7, and in step (2), below, for the Computing Sight M7A1.

(1) ALINEMENT OF TELESCOPES, COMPUTING SIGHT M7.

*(a)* Place the bore sights in the gun.

*(b)* Remove the wing screw which connects the link from the trunnion to the superelevation cam shaft arm.

*(c)* Tip the deflection mechanism control case forward by rotating the superelevation cam shaft arm.

*(d)* Remove the bore sighting plug from its clip and insert it in the hole in the superelevation cam arm, alining this hole with the hole in the superelevation cam link. See figure 166 of the sight in the bore sighting position.

*(e)* Set the speed to zero by turning the speed control handwheel.

*(f)* Bore sight the gun on a celestial body or other distant object.

*(g)* Adjust the telescope holders laterally or vertically as necessary to bring the center of the vertical line of the Azimuth Telescope M7 and the center of the "A-A" line of the Elevation Telescope M74, to on target position.

*(h)* Pull out the bore sighting plug and place it back in its clip. Tip the deflection mechanism control case into its normal vertical position and connect the link from the trunnion to the superelevation cam arm with its wing screw.

(2) ALINEMENT OF TELESCOPE, COMPUTING SIGHT M7A1.

*(a)* Place the bore sights in the gun.

*(b)* Remove the wing screw which locks the superelevation cam arm and the superelevation arm yoke together.

*(c)* Tip the deflection mechanism control case forward by rotating the superelevation cam shaft arm.

*(d)* Remove the bore sighting plug from its clip and insert it in the hole in the superelevation cam arm, alining this hole with the hole in the superelevation cam link. See figure 166 of the sight in the bore sighting position.

*(e)* Set the speed to zero by turning the speed control handwheel.

*(f)* Bore sight the gun on a celestial body or other distant object.

*(g)* Adjust the telescope holders laterally or vertically as neces-

## 40-MM AUTOMATIC GUN M1 (AA) AND 40-MM ANTIAIRCRAFT GUN CARRIAGES M2 AND M2A1

sary to bring the center of the vertical line of the azimuth Telescope M7 and the center of the "A-A" line of the elevation Telescope M74 to on target position.

(h) Pull out the bore sighting plug and place it back in its clip. Tip the deflection mechanism control case into its normal vertical position and replace the wing screw which locks the superelevation cam arm and the superelevation cam arm yoke together.

c. **Verification and Adjustment of the Vertical Axis of Deflection Mechanism, Computing Sight M7.**

(1) Level the gun carriage.

(2) Bring the gun tube to zero degree elevation.

(3) Place a gunner's quadrant set to zero on the milled surfaces of the deflection mechanism control housing using parallel bar part No. 7578746.

(4) Adjust the length of the link between the superelevation cam arm and the trunnion until the deflection mechanism control housing is accurately leveled as shown by the gunner's quadrant.

(5) Elevate the gun to 0 degree −15 degrees −30 degrees −45 degrees −60 degrees −75 degrees. The deflection mechanism control housing should remain level within $\pm 2$ mils. If the deflection mechanism control housing does not remain level within $\pm 2$ mils, perform the adjustments as indicated below.

(6) If the deflection mechanism control housing tips forward or backward more than 2 mils when performing the check in step (5), above, it may be corrected in the following manner:

(a) If the deflection mechanism control housing tips forward, it means that the pivot at the trunnion is either too low and the link between the trunnion and the superelevation cam arm is slightly too long, and/or too far back and the link is too long by an equal amount. In order to correct the first of these conditions, move the pivot up in small steps by its eccentric and shorten the link slightly. Check at 0 degree −15 degrees −30 degrees −45 degrees −60 degrees −75 degrees after each adjustment. If this does not correct the error, correct for the second condition. In order to correct for the second condition, move the pivot forward in small steps by its eccentric and shorten the link in equal steps. Check after each adjustment of 0 degree −15 degrees −30 degrees −45 degrees −60 degrees −75 degrees.

(b) If the deflection mechanism control housing tips backward, it means that the pivot is either too high and the link is slightly too long, and/or too far forward and the link is too short by an equal amount. In order to correct this first condition, lower the pivot in small steps by its eccentric and shorten the link slightly. Check after each adjustment of 0 degree −15 degrees −30 degrees −45 degrees −60 degrees −75 degrees. If this does not correct the error, correct for

## SIGHTING AND FIRE CONTROL EQUIPMENT

the second condition. In order to correct this second condition, move the pivot back in small steps by its eccentric and lengthen the link in equal amounts. Check after each adjustment of 0 degree −15 degrees −30 degrees −45 degrees −60 degrees −75 degrees.

(c) When the above adjustments have been made, make certain all nuts, screws, and bolts used in making the adjustments are secure.

d. **Verification and Adjustment of the Vertical Axis of Deflection Mechanism—Computing Sight M7A1.**

(1) Level the gun carriage.

(2) Bring the gun tube to zero degree elevation.

(3) Place a gunner's quadrant set to zero on the milled surfaces of the deflection mechanism control housing using parallel bar, Part No. 7578746.

(4) Adjust the length of the link between the superelevation cam arm and the trunnion until the deflection mechanism control housing is accurately leveled as shown by the gunner's quadrant.

(5) Elevate the gun to 0 degree −15 degrees −30 degrees −45 degrees −60 degrees −75 degrees. The deflection mechanism control housing should remain level within $\pm 2$ mils. If the deflection mechanism control housing does not remain level within $\pm 2$ mils, perform the adjustments as indicated below.

(6) If the deflection mechanism control housing tips forward or backward more than 2 mils when performing the check in step (5), above, it may be corrected in the following manner:

(a) If the deflection mechanism tips forward, it means that the link pivot on the superelevation cam arm is either too high and the link connecting the superelevation cam arm to the trunnion is slightly too long, and/or too far toward the rear of the gun and the link is too short by an equal amount. In order to correct the first of these conditions, move the pivot down by means of its eccentric in small steps and shorten the link slightly. Check at 0 degree −15 degrees −30 degrees −45 degrees −60 degrees −75 degrees after each adjustment. If this does not correct the error, correct for the second condition. In order to correct for the second condition, move the pivot forward in small steps by means of its eccentric and lengthen the rod by equal amounts. Check at 0 degree −15 degrees −30 degrees −45 degrees −60 degrees −75 degrees after each adjustment.

(b) If the deflection mechanism tips backwards, it means that the pivot is either too low and the link is slightly too long and/or too far forward and the rod is too long by an equal amount. In order to correct for the first of these conditions, move the pivot up in small steps by its eccentric and shorten the link slightly. Check after each adjustment of 0 degree −15 degrees −30 degrees −45 degrees −60 degrees −75 degrees. If this does not correct the error, correct for

TM 9-252
93-94

**40-MM AUTOMATIC GUN M1 (AA) AND 40-MM ANTIAIRCRAFT GUN CARRIAGES M2 AND M2A1**

the second condition. In order to correct for the second condition, move the pivot back in small steps by means of its eccentric and shorten the link in equal amounts. Check over each adjustment at 0 degree −15 degrees −30 degrees −45 degrees −60 degrees −75 degrees.

(c) When the above adjustments have been made, make certain all nuts, screws, and bolts used in making the adjustments are secure.

e. Verification of Alinement of Main Assembly With Bore.
(1) Level the gun carriage.
(2) Level the gun to zero elevation, using a gunner's quadrant set at "0."
(3) Test level of the main bracket on the azimuth side by putting the gunner's quadrant across its two bosses. If there is a deviation of over 5 mils above or below zero elevation, report to the responsible ordnance unit.

f. Verification of Azimuth Drive to Arrow.
(1) Level the gun carriage.
(2) Level gun to approximately zero elevation and aline bore and axis of arrow shaft with a distant object.
(3) Set speed to 500 miles.
(4) Traverse gun through 360 degrees. If axis of arrow does not remain directed at the same object, report to the responsible ordnance unit.

**94. CARE AND PRESERVATION, COMPUTING SIGHTS M7 AND M7A1.**

a. General Precautions.
(1) Disassembly and assembly by the using arms is permitted only to the extent specifically authorized herein. Turning of screws or other parts not incident to bore sighting, alinement of telescopes, or to the use of the system is expressly forbidden.
(2) Keep the system clean and in condition for traveling when not in use, the telescopes in their cases and the canvas covers in place.

b. Lubrication.
(1) Ball bearings, gear drives, covered joints and the flexible shaft are lubricated at assembly. Further lubrication of these will be required only at long intervals, and is performed by ordnance personnel.
(2) Telescope yoke surfaces, the telescoping rod to the azimuth telescope yoke, and link clevis pins should be frequently oiled with OIL, lubricating, for aircraft instruments and machine guns.
(3) Keep the outside of the flexible shaft casing free of oil.

c. Lighting Devices. Remove the flashlight battery cells whenever the lighting devices are not in use for several days. Chemical re-

## SIGHTING AND FIRE CONTROL EQUIPMENT

action set up in the cells as they become exhausted will cause the cells to swell, thereby making removal difficult and damaging the battery tube.

d. Telescopes.

(1) To obtain satisfactory vision, it is necessary that the exposed surfaces of the lenses be kept clean and dry. Corrosion and etching of the surface of the glass which greatly interfere with the good optical qualities of the instrument can be prevented or greatly retarded by keeping the glass clean and dry.

(2) Under no conditions should polishing liquid, paste, or abrasives be used for polishing lenses.

(3) For wiping optical parts, use only PAPER, lens, tissue. Use of cleaning cloth on optical glass is not permitted. To remove dust, brush the glass lightly with a clean, camel's-hair brush and rap the brush against a hard body to knock out the small particles of dust that cling to the hairs. Repeat this operation until all dust is removed.

(4) Exercise particular care to keep the lens free from all grease. Do not wipe the lens with the fingers. To remove a slight amount of grease from the lenses, breathe heavily on the glass and wipe off with PAPER, lens, tissue; repeat this operation several times until clean. If inspection reveals that the internal optics are dirty or greasy, the telescopes must be turned over to the nearest ordnance unit.

(5) Moisture due to condensation may collect on the optical parts of the telescopes when the temperature of the parts is lower than that of the surrounding air. This moisture, if not excessive, can be removed by warming the telescopes in a warm place. Heat from strongly concentrated sources should never be applied directly, as it may cause unequal expansion of parts, resulting in breakage of lens, or inaccuracies in observation.

## 95. BORE SIGHT.

a. General. The bore sight is used to indicate the direction of the axis of the bore of the gun, for alinement and verification of sights. Each bore sight is composed of a breech element and muzzle element.

b. Description.

(1) The breech bore sight is a hollow cylinder with one end closed except for a small aperture. This sight fits accurately into the breech chamber of the gun.

(2) The muzzle bore sight consists of a tube with an extension at one end retained by a chain (to which is attached a disk with an aperture near the outer edge) and at the other end of the tube two cross bars are welded.

c. Operation.

(1) Remove the section of the cartridge case deflector which is

**40-MM AUTOMATIC GUN M1 (AA) AND 40-MM ANTIAIRCRAFT GUN CARRIAGES M2 AND M2A1**

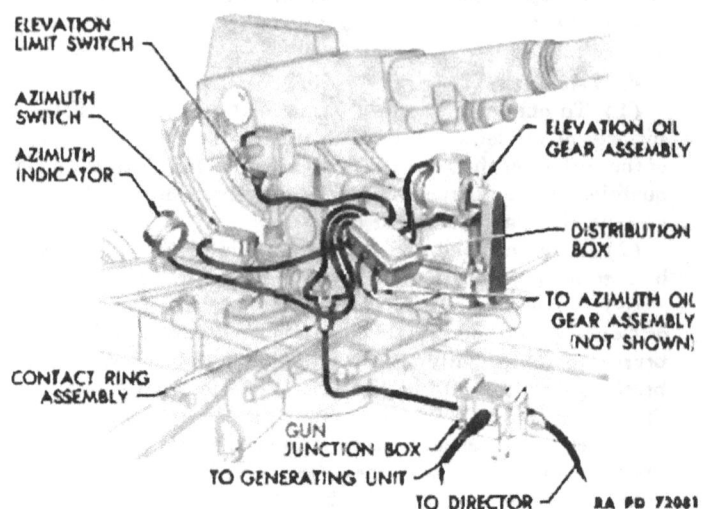

*Figure 167 — Remote Control System M5 — On-carriage Components*

attached to the rear of the breech casing. Place the breech bore sight in the breech opening. Attach the muzzle bore sight.

(2) Look through the aperture in the breech bore sight, the direction of the axis of the bore being indicated by the intersection of the cords at the muzzle.

**96. DESCRIPTION OF REMOTE CONTROL SYSTEM M5.**

a. General.

(1) The remote control system (fig. 167) is an electrically controlled hydraulic power system designed for use with the 40-mm Gun Carriages M2 and M2A1 (AA). It is controlled by the AA Director M5A2, M5A1, or M5. Power is supplied from the Generating Unit M5. The function of the remote control system as a whole is to point the gun accurately in azimuth and elevation according to the output data supplied from the AA director. This type of control is known as "remote gun control."

(2) To oil gear units, one for azimuth and one for elevation, supply controlled driving power for traversing and elevating the gun. The gun must be brought into approximate alinement with the director before the oil gear may be engaged but, once the oil gear is engaged, the gun will follow the director accurately. Approximate alinement of the gun with the director is shown by an azimuth indicator toward

## SIGHTING AND FIRE CONTROL EQUIPMENT

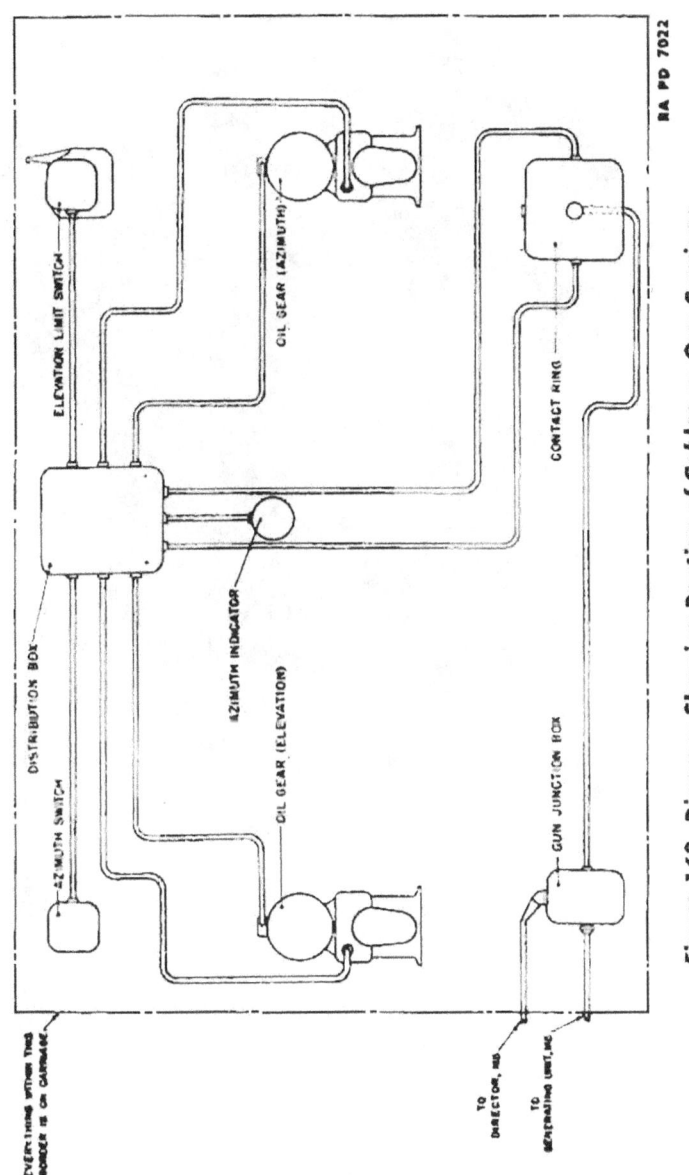

*Figure 168 — Diagram Showing Routing of Cables on Gun Carriage*

Figure 169 — Gun Junction Box

# SIGHTING AND FIRE CONTROL EQUIPMENT

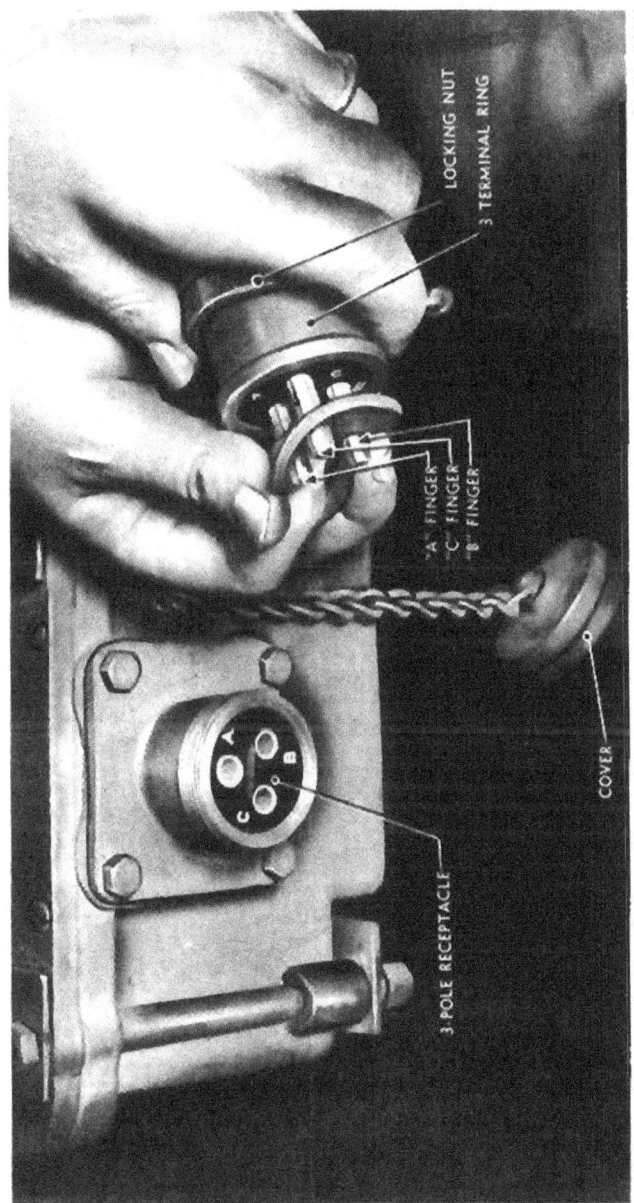

Figure 170 — 3-hole Receptacle and Matching Fingers of Cable Plug

**TM 9-252**
**96**

**40-MM AUTOMATIC GUN M1 (AA) AND 40-MM ANTIAIRCRAFT GUN CARRIAGES M2 AND M2A1**

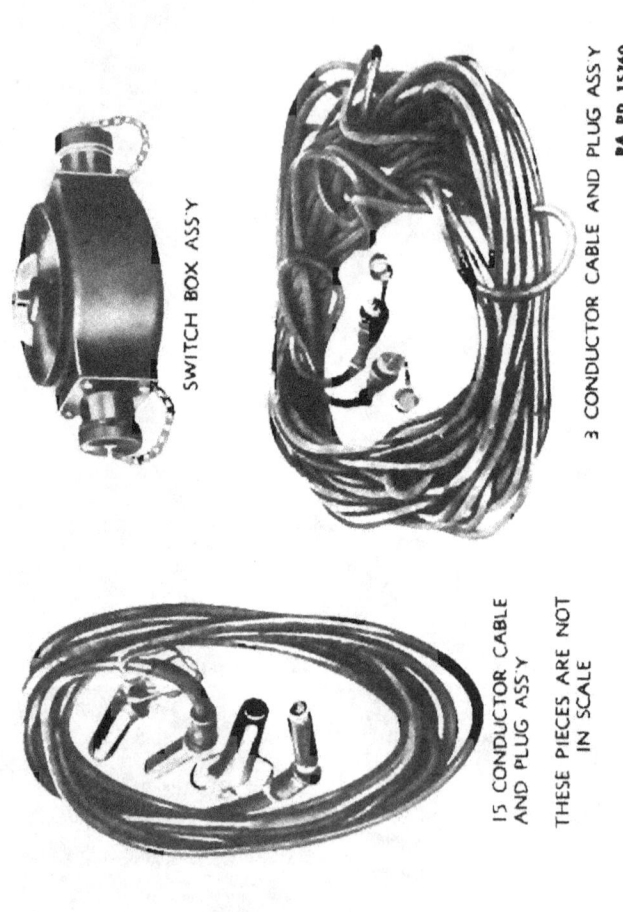

*Figure 171 — Cable System M8 — (15-foot Cable Not Shown)*

## SIGHTING AND FIRE CONTROL EQUIPMENT

Figure 172 — Oil Gear—Remote Control System M5

**TM 9-252**
**96**

**40-MM AUTOMATIC GUN M1 (AA) AND 40-MM ANTIAIRCRAFT GUN CARRIAGES M2 AND M2A1**

Figure 173 — Elevation Limit Switch

## SIGHTING AND FIRE CONTROL EQUIPMENT

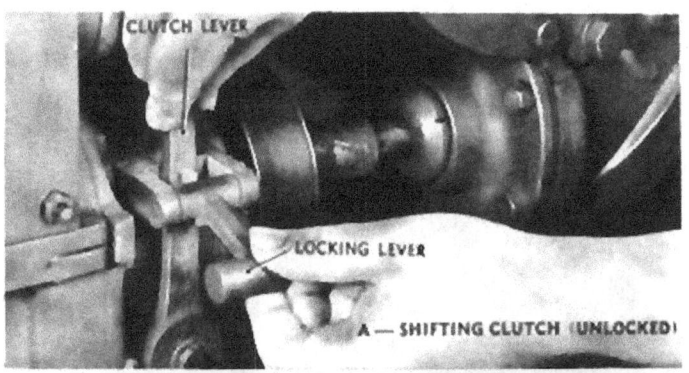

*Figure 174 — Elevation Oil Gear Clutch Lever Showing Operation of Locking Lever*

TM 9-252
96

**40-MM AUTOMATIC GUN M1 (AA) AND 40-MM ANTIAIRCRAFT GUN CARRIAGES M2 AND M2A1**

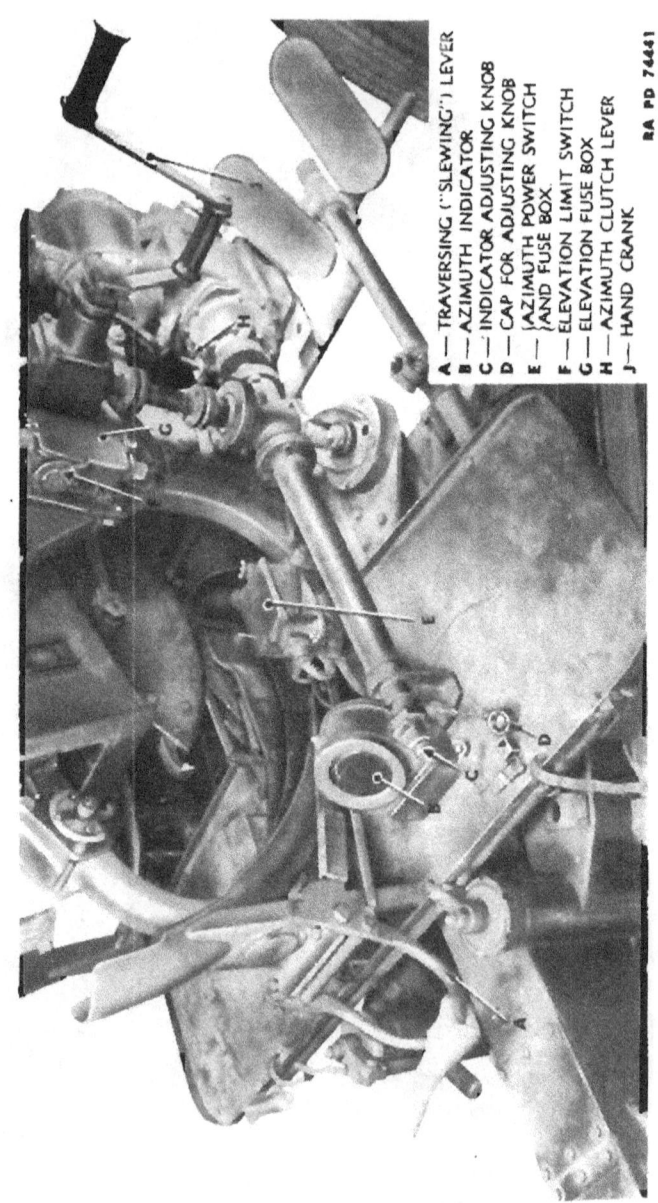

Figure 175 — Lowering the Power Synchronizing Handle Turns "ON" Azimuth Power Switch and Engages Azimuth

A — TRAVERSING ("SLEWING") LEVER
B — AZIMUTH INDICATOR
C — INDICATOR ADJUSTING KNOB
D — CAP FOR ADJUSTING KNOB
E — AZIMUTH POWER SWITCH AND FUSE BOX
F — ELEVATION LIMIT SWITCH
G — ELEVATION FUSE BOX
H — AZIMUTH CLUTCH LEVER
J — HAND CRANK

RA PD 74441

## SIGHTING AND FIRE CONTROL EQUIPMENT

the rear of the loading platform near the power synchronizing mechanism.

(3) The system permits continuous tracking in azimuth, but is limited in elevation by an elevation limit switch which automatically goes to "OFF" before the mechanical limit stops are reached.

b. **Cable Connections.** The arrangement and approximate location of the principal on-carriage parts of the remote control system are shown in figure 167. The cable connections are shown in figure 168. Cable connections from the director and generating unit are made to the gun junction box (fig. 169). The cable from the director is 30 feet (or 60 feet) long, and the cables from the generating unit are 225 feet long and 15 feet long (Cable System M8, fig. 171), joined by the switch box. Electrical connections pass to the top carriage through the contact ring, which permits traversing the gun without kinking or twisting the external connecting cables.

c. **Gun Junction Box.** The gun junction box (fig. 169) is the unit to which the cables from the director and generating unit are connected. When not in use, a cover protects the receptacles. One receptacle has three holes marked "A," "B," and "C" to receive the generating unit cable (fig. 170). The fingers of the cable plug are correspondingly marked. The other receptacle accommodates the director cable which has a "D" plug with fifteen terminal rings (fig. 171).

d. **Oil Gears.** Two oil gear units are mounted on the carriage (fig. 167). The elevation oil gear is on the left side, and the azimuth oil gear is on the right side. Each oil gear (fig. 172) is a weathertight assembly consisting of an electric motor at the top, then an electrical control unit and, at the bottom, an oil pump and oil motor in a single housing. The electric motor drives the oil pump through a chain drive which is inclosed in an oiltight housing. An arrow on the motor housing indicates the direction in which chain should turn. The electrical control unit operates a pilot valve (fig. 183) which controls the flow of oil from the oil pump to the oil motor.

e. **Elevation Limit Switch.**

(1) Power to the elevation oil gear is controlled by the elevation limit switch (fig. 173).

(2) When the switch is turned "ON" (B, fig. 173) the elevation oil gears are energized.

(3) When the switch is turned "OFF" (A, fig. 173) (either by hand or automatically), the elevation oil gear does not operate. The gun can be operated manually by engaging the hand cranks.

(4) When operating by remote control from the director, the elevation limit switch trips and is automatically turned "OFF" at the upper and lower limits of elevation.

(5) When the elevation limit switch cuts off automatically, it is

## 40-MM AUTOMATIC GUN M1 (AA) AND 40-MM ANTIAIRCRAFT GUN CARRIAGES M2 AND M2A1

*Figure 176 — Azimuth Clutch Lever Disengaged by Blocking (For Adjustment Purposes Only)*

necessary to engage the hand cranks, elevate or depress the gun slightly, and then turn "ON" the elevation limit switch by hand (B, fig. 173).

(6) The upper limit is normally fixed at 85 degrees, while the lower limit is fixed at a point depending upon the emplacement of the gun so that the line of fire is above any intervening obstruction. For adjustment of these limits, see paragraph 98 h.

f. **Elevation Clutch Lever.**

(1) The lower rear portion of the oil gear houses a clutch mechanism for disengaging or engaging the oil gear. The elevation oil gear clutch (fig. 174) is engaged or disengaged by hand. It has an auxiliary locking lever to lock the elevation clutch lever in the "IN" or "OUT" positions.

(2) Whenever it is desired to shift the clutch lever from the "IN" or "OUT" positions, the locking lever is lifted to unlock and enable the clutch lever to be shifted, and must be lowered to lock the clutch lever after it is shifted (A, fig. 174).

(3) The elevation oil gear is engaged ("IN") when the top of the lever is in the forward position (B, fig. 174). In this "IN" position, the elevation oil gear is using hydraulic pressure to elevate the gun automatically by remote control from the director.

## SIGHTING AND FIRE CONTROL EQUIPMENT

*Figure 177 — Azimuth Indicator — "Match-the-pointer" Dial*

*Figure 178 — Azimuth Indicator — Modified "Red-black-white" Dial*

(4) The elevation oil gear is disengaged ("OUT") when the top of the clutch lever is in the rear position (C, fig. 174). In this "OUT" position, the gun is free from director remote control and is operated by the elevation handwheel at the gun.

CAUTION: The elevation limit switch must be "OFF" before engaging or disengaging elevation clutch.

g. Azimuth Clutch and Power Switch.

(1) The azimuth power switch is automatically turned "ON" when the power synchronizing ("SLEWING") handle is depressed. This energizes the azimuth oil gear.

(2) When this handle is depressed (fig. 175), the azimuth oil gear clutch is also automatically engaged ("IN"). The gun now operates on remote control from the director.

(3) When the power synchronizing handle is raised, the azimuth switch cuts "OFF" automatically, and the clutch is automatically thrown "OUT" (disengaged). The gun then does not operate by remote control and is slewed by hand.

(4) If it is desired to cut off the azimuth switch while the power synchronizing handle is depressed (for adjustment purposes), it is necessary to block the switch as illustrated in figure 176.

h. Azimuth Indicator.

(1) The azimuth indicator (figs. 177 and 178) is used as an alinement indicator when synchronizing the gun with the director, after slewing the top carriage.

(2) The "match-the-pointer" indicator (fig. 177) consists of a synchronous repeater, containing a circular black shutter with a white pointer which follows the rotation of the coarse azimuth transmitter

TM 9-252
96

**40-MM AUTOMATIC GUN M1 (AA) AND 40-MM ANTIAIRCRAFT GUN CARRIAGES M2 AND M2A1**

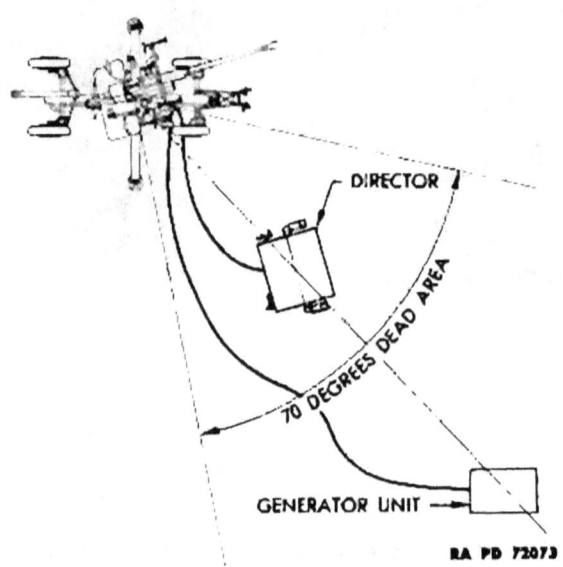

*Figure 179 — Diagram Illustrating "Dead Sector"*

in the director, and a mechanical dial with a white pointer which is geared to the traversing mechanism of the gun carriage.

(3) After the gun has been slewed by hand, its synchronization with the director is disturbed (it being either behind or ahead of the director). When the carriage is within 10 degrees of alinement with the director, the indicator shows this by means of two limit lines. Within these limits, the electrical system will automatically synchronize the gun and director without attention of the gun detail. If the gun is out of alinement by more than 10 degrees, the electrical system cannot synchronize the gun with the director and the gun will be "out of phase." The indicator shows this by the matching pointer being beyond the limit marks (fig. 177) and the gun must be slewed until the indexes match approximately.

(4) The "red-black-white" (blackout) indicator may be found in some guns. This indicator has red, black, and white sections in the dial. The matching position is indicated when the dial is "blacked out" (or almost blacked out). If the gun is out of phase with the director, either the red or the white markers will show. A narrow strip of red or white indicates satisfactory alinement.

(5) The majority of the "red-black-white" (blackout) indicators

## SIGHTING AND FIRE CONTROL EQUIPMENT

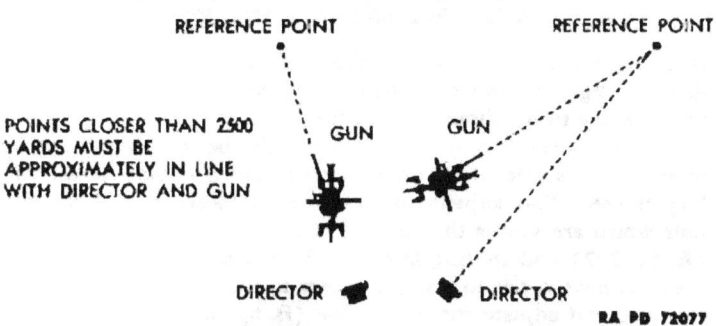

Figure 180 — Choosing a Reference Point

have been modified. The dials have been painted black except for the alining marks, one on the mechanical dial and one on the shutter (fig. 178). This change makes the indicator similar to the standard indicator.

(6) INDICATOR ADJUSTING KNOB. When the gun and director are alined, the indicator marks should match exactly. Otherwise, the indicator should be adjusted. Remove the adjusting knob cover (fig. 183), depress, and turn the knob until the "match-the-pointer" indicator marks match (fig. 177) or the "blackout dial" blacks out.

## 97. SETTING UP, ORIENTING, AND OPERATION, REMOTE CONTROL SYSTEM M5.

a. Initial Set-up. Before firing operations are started, the following steps should be taken:

(1) Level the gun carriage accurately. Instructions for leveling the gun carriage are contained in paragraph 37.

(2) Orient the director with the gun as instructed in paragraph 94 c.

(3) Connect the generating unit and director to the gun mount by the cables provided, making sure that all switches are "OFF" until cable connections have been completed. Turn the "D" plug of the director to lock the plug in its socket of the gun junction box (fig. 169). The plugs of the generating unit cable have a round nut which must be screwed to the body of the mating receptacle of the gun junction box.

(4) The main power switch is on the control panel of the generating unit. This switch should not be turned "ON" until the cables have been connected and the cable plugs secured.

b. Emplacing and Setting Up.

(1) Since there are no provisions for parallax corrections, and since in the prescribed method of fire adjustment it is necessary for the trackers to see tracers that cross the gun target line, emplace the director 13 to 15 feet from the pintle center of the gun. This results

## 40-MM AUTOMATIC GUN M1 (AA) AND 40-MM ANTIAIRCRAFT GUN CARRIAGES M2 AND M2A1

in a dead sector which extends about 35 degrees to either side of the director (fig. 179) in which the gun should ordinarily not be fired, to avoid injury to the director personnel.

(2) The elevation limit switch should be set to cut off automatically at a safe lower limit to clear any obstructions within the line of fire. This adjustment is made by loosening the two clamp nuts which are visible through the slot of the rotating trunnion plate (A, fig. 187) and shifting laterally the stops. If the trunnion plate does not have a slot to provide access to the nuts, the plate must be removed and adjustment then made (B, fig. 187).

c. Orienting.

(1) GENERAL.

(a) Orienting is the alining of the director and gun on a reference point in azimuth and elevation. This is necessary so that the gun will point in the direction indicated by the director (as illustrated in figure 180).

(b) A clearly defined object, preferably in the field of fire and more than 2,500 yards distant, should be selected as a reference point. If an object 2,500 yards distant cannot be obtained, a closer point may be used if it is approximately in line with the director and gun.

(c) Set up the gun and director according to subparagraph h, above.

(d) Level the gun and director.

(e) Be sure that all switches are off at the gun and at the director.

(f) Set zero deflection in the director by pushing in the rate setting clutch knob and turning the tracking handwheel until the white mark on the deflection gear is lined up with its index marker. Then disengage the rate setting clutch by pulling the knob out. Deflection will be zero in both azimuth and elevation.

(g) Connect the generator cable to the gun *only*, and turn on the master switch at the generator.

(h) Turn on the power to the azimuth oil gear by depressing the power synchronizing handle.

(i) Check the dither of the azimuth oil gear and adjust if necessary (par. 98 d).

(j) Adjust the azimuth oil gear to neutral (par. 98 e).

(k) Place bore sights on the gun.

(l) Turn on the power to the elevation oil gear by depressing the elevation limit switch. (Be sure the clutch is engaged.)

(m) Check the dither of the elevation oil gear and adjust if necessary (par. 98 d).

(n) Adjust the elevation oil gear to neutral (par. 98 e).

(o) Shut off the power at the main power switch of the generator and connect the gun and director through the fifteen-conductor cable.

## SIGHTING AND FIRE CONTROL EQUIPMENT

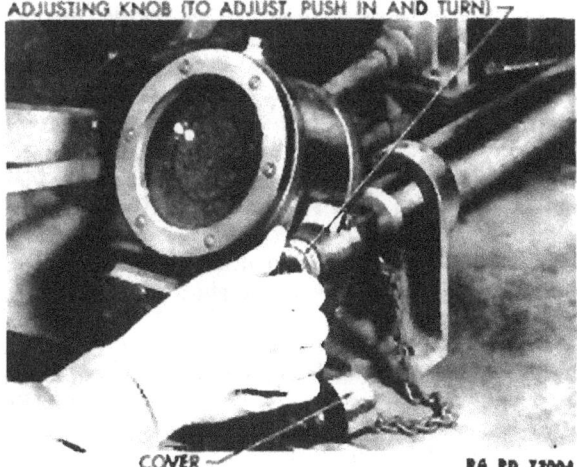

*Figure 181 — Making Azimuth Indicator Adjustment*

(p) Turn on the master switch at the generator.

(2) ORIENTING IN AZIMUTH.

(a) Turn on the power at the director and gun (azimuth side only).

(b) Bore sight the gun on the previously selected orienting point, moving the gun by means of the director.

(c) Disengage the orienting clutch in the following manner:

1. Remove the orienting clutch cover.
2. Tighten the small locking nut.

NOTE: This step must be performed first; otherwise, the gun will move when the large knob is loosened.

3. Loosen the large knob.
4. Replace the orienting clutch cover.

(d) Traverse the director with the azimuth handwheel until the director telescopes are on the orienting point.

(e) Re-engage the orienting clutch by reversing the procedure given in step (c), above.

(f) Traverse the director several times and see if the gun and director aline on the orienting point.

(g) Observe the azimuth indicator. The two white markers should be alined (or the dial blacked out) (fig. 181).

(h) The gun is now oriented in azimuth.

(3) ORIENTING IN ELEVATION.

(a) Set the elevation dials of the director at a convenient reading, for example 45 degrees.

221

## 40-MM AUTOMATIC GUN M1 (AA) AND 40-MM ANTIAIRCRAFT GUN CARRIAGES M2 AND M2A1

(b) With the elevation limit switch off, elevate the gun manually to 45 degrees (to correspond to the reading on the director).

(c) Observe the elevation scale plate on the gun for a proper reading.

(d) Turn on the power to the elevation oil gear by depressing the elevation limit switch.

(e) Check the position of the gun with a gunner's quadrant.

(f) The gun should point at 45 degrees. If not, make adjustments on the resetter gear train of the elevation oil gear.

(g) Traverse the gun in elevation several times.

(h) Recheck the alinement of the gun and director at 45 degrees with the gunner's quadrant.

(i) The gun is oriented in elevation.

PRECAUTION: Be particularly careful when orienting in elevation that the elevation motor switch at the director is "OFF." Otherwise, a superelevation will be set in, and orienting will be in error by the amount of superelevation.

d. Tracking the Target.

(1) Turn on the main switch on the generating unit to energize the system.

(2) Check that the elevation clutch lever is in the "IN" position (oil gear engaged).

(3) Turn on the elevation limit switch on the gun carriage to energize the elevation oil gear. (The elevation oil gear clutch should always be engaged before the elevation switch is turned on.)

(4) When the director picks up the target, bring the gun into line with the director by traversing the gun until the azimuth indicator pointers match (or blacks out). It is not necessary for the azimuth indicator to match exactly (or black out exactly) when bringing the gun into line, as the gun will fall into line with the director if it is initially alined within 10 degrees.

(5) Now depress the power synchronizing ("SLEWING") handle. This operation throws the azimuth switch and engages the clutch in the azimuth oil gear.

(6) After the director picks up and tracks the target, it computes the firing azimuth and quadrant elevation and transmits these electrically to the oil gears which operate the gun. All human pointing errors at the gun are eliminated by the remote control system. Once the target is picked up and a rate set in, no correction can be made at the gun.

(7) Whenever the gun has engaged one of the elevation limit stops and tripped the elevation limit switch, it will be necessary to elevate or depress the gun manually by engaging the hand crank until it is brought inside the cut-off limit, and realined with the director.

TM 9-252
97

## SIGHTING AND FIRE CONTROL EQUIPMENT

As soon as this has been done, disengage the hand crank and reset the elevation limit switch.

*e. Changing Targets.*

(1) When the order is given to track a new target, lift the power synchronizing handle and slew the gun to aline with the director again. When the azimuth indicator matches (or "blacks out"), indicating alinement with the director, depress the lever to engage the clutch.

(2) The elevation clutch and switch should remain engaged throughout active operation, as there is no provision for slewing the gun in elevation.

*f. Cease Tracking.* Both gun and the director should be set at an elevation of 30 degrees (or any other selected elevation, provided the same setting is used at both the gun and director) so that a new target can be picked up without delay.

*g. Operating Faults.*

(1) If the gun does not follow the director both in azimuth and elevation when tracking a target, the errors made are caused either by the mistakes of the operators at the director and the gun or by some mechanical or electrical imperfections in the remote control system or the director. Operating faults at the gun may be the following:

*(a) Bad Lining-up.* The breech peep sight and muzzle cross bars should be used, if they are available (par. 95).

*(b) Gun Out of Level.* The correct leveling of the gun cannot be overstressed. Any error in leveling will cause a corresponding error in elevation.

*(c)* Depressing the power synchronizing handle before the gun is approximately in line with the director (within 10 degrees).

*(d)* Turning on limit switch before gun is approximately in line with the director (within 10 degrees).

*(e)* Forgetting to reset the elevation limit switch after it has been automatically thrown to "OFF."

(2) If the director is operated at maximum rate and with maximum range it is possible for gun to "jump phase" in elevation, that is, synchronize within the next 20-degree sector.

(3) In cold weather, if oil gears are not properly warmed up before tracking with director, a similar condition as noted above can also occur.

*h. Precautions.* The following precautions should be taken in the operation and maintenance of the system:

(1) Power should be switched off before cables are connected or disconnected. See that cables are securely held in the receptacles before turning the power on.

223

TM 9-252
97-98

**40-MM AUTOMATIC GUN M1 (AA) AND 40-MM ANTIAIRCRAFT GUN CARRIAGES M2 AND M2A1**

*Figure 182 — Indicators on Clutch Housing and Output Coupling for "Accuracy Test"*

(2) The elevation clutch lever should be put in the "IN" position (top of lever away from the coupling) before switching on the power supply.

(3) Never put oil from an unsealed container in the oil gears.

(4) Be sure that the oil gears have been filled with oil correctly as noted in paragraph 102.

(5) Be sure the carriage is level before firing. The correct leveling of the carriage cannot be overstressed.

(6) In orienting for elevation or in adjusting oil gears to neutral, the unit cover plates should not be removed when dust or rain can get into unit, unless unavoidable. If it is necessary to remove covers under adverse conditions, the units should be protected to insure that no dust or rain gets into the unit.

i. **Practice Tracking.** To avoid damage to the electrical units when practicing with the director, energize the oil gears at the gun, and practice with the gun under direct control, or disengage the clutch on the elevation oil gear and turn on power on that gear, and throw in the power synchronizing handle at the gun (in order to protect the azimuth oil gear and the azimuth indicator unit).

**98. TESTS AND ADJUSTMENTS OF REMOTE CONTROL SYSTEM M5.**

a. **General.** The following described tests and adjustments must be made at regular intervals in order to insure proper operation of the system. All of these tests and corrective adjustments are made "on-carriage."

## SIGHTING AND FIRE CONTROL EQUIPMENT

*b. Accuracy Test and Adjustment.*

(1) The purpose of this test is to check over-all operation of system. Place a reference mark or indicator on the clutch housing and a corresponding mark made on circumference of output coupling (fig. 182). Traverse the director exactly 6 degrees (14 degrees for checking elevation unit), making sure that director does not overrun reading. The output coupling should make one complete revolution $\pm \frac{1}{16}$ inch (3 revolutions $\pm \frac{3}{32}$ inch for elevation). Test should be run in both directions.

(2) If unit does not meet accuracy test requirement, the following operations are required:

(a) Adjust dither (subpar. d, below).
(b) Make neutral ("creep") adjustment (subpar. e, below).
(c) Add oil to make up any deficiency (par. 102).
(d) Make backlash test and adjustment of transmitter gearing (subpar. c, below).

*c. Adjustment of Backlash in Transmitter Gearing.*

(1) Remove cover plate over transmitter units.

(2) Check for backlash or play between the teeth of the spiral gear and the pinion.

(3) If adjustment is necessary, loosen the screws holding the pinion shaft bearing bracket (fig. 183).

(4) Press the bracket upward, adjusting so that the pressure is as light as possible while there is no play between the gears.

(5) Reclamp the screws and test again to be sure that the gears run together smoothly and freely after adjustment.

CAUTION: This adjustment should be made only by qualified battery mechanic.

*d. Dither Adjustment.*

(1) The dither mechanism is such that it sets up a vibration in the oil gear which can be felt by placing the hand on the output coupling of the oil gear and twisting slightly to take up the slack. Adjustment of the dither is made as follows:

(a) To the left side of the oil gear on the case appears the word "DITHER" (A, fig. 184).

(b) Remove the hexagonal cap.

(c) With a screwdriver, turn the adjusting screw (B, fig. 184) until a definitely perceptible tremble is felt on the output coupling of the oil gear. Turn to the left to increase, to the right to decrease the vibration.

(2) Proper adjustment of the dither is best made by a sense of feel, but generally the correct adjustment is such that it will not cause the gun tube to tremble but will give a vibration to the output coupling of the oil gear which is just barely visible. Insufficient dither results in an inaccurate and insensitive gun, and excessive dither

TM 9-252
98

**40-MM AUTOMATIC GUN M1 (AA) AND 40-MM ANTIAIRCRAFT GUN CARRIAGES M2 AND M2A1**

Figure 183 — Transmitter Spiral Gear and Pinion and Pilot Valve

## SIGHTING AND FIRE CONTROL EQUIPMENT

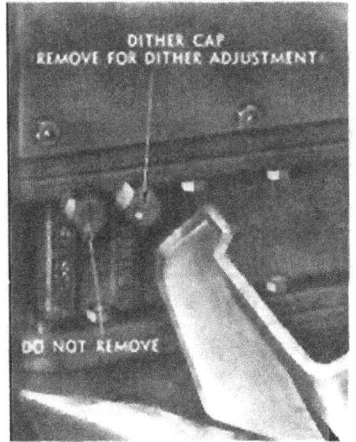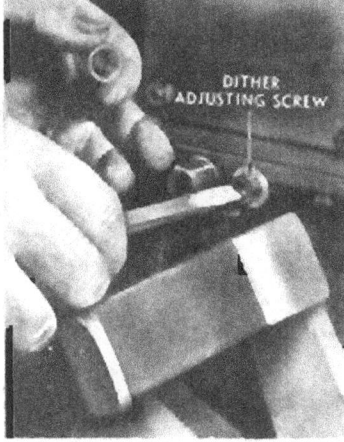

*Figure 184 — Dither-Adjustment*

may cause low top speed, rough tracking, or slightly erratic action. It is necessary to have the oil gear thoroughly warmed up before making the adjustment, as the dither is noticeably affected by temperature.

(3) After making the dither adjustment, always make the neutral ("creep") test (subpar. e, below), as adjustment of the dither may cause oil gear to creep again by disturbing the pilot valve adjustment.

CAUTION: Do not turn dither adjusting screw more than several turns, as otherwise the mechanism may become disengaged.

e. Neutral ("Creep") Adjustment.

(1) When the director is not tracking, the pilot valve should be in neutral position so that there is no turning (or "creep") of the oil gear output coupling.

(2) With the oil gear clutch engaged, switch power off at the generating unit, and remove director cable from gun junction box. It is imperative that the director be disconnected from the gun while testing for, and while making neutral (creep) adjustment. This will make the electrical differential "electrically dead" and allow springs to hold torque arm (fig. 183) in neutral position. Turn power on.

(3) Observe the output coupling of oil gear unit and notice if there is any rotation (creep). If there is any creep, time it. Maximum allowance for creep is one revolution in 2 minutes. If creep is more than this rate, neutral (creep) adjustment should be made.

(4) Make adjustment by turning the adjusting thumbscrew (A, insert, fig. 185) until the creep of the output coupling is removed.

(5) Move the differential torque arm up and down slightly and

## 40-MM AUTOMATIC GUN M1 (AA) AND 40-MM ANTIAIRCRAFT GUN CARRIAGES M2 AND M2A1

*Figure 185 — Making the Neutral ("Creep") Adjustment*

then allow it to spring back into neutral position. Repeat several times in alternate direction, observing whether the output coupling comes to rest without creep after each release, repeating the neutral (creep) adjustment if necessary, until creep has been satisfactorily removed.

(6) It is possible that the pilot valve may become stuck and operation described in step (5), above, will not free it. In that case, remove the pilot valve (as instructed in par. 100 a), clean with SOLVENT, dry-cleaning, and replace (as instructed in par. 100 b).

f. *Adjustment of Resetter Spiral Gear.*

(1) If, after making orienting operations described in paragraph 97 c, the gun and director are not oriented in elevation, an adjustment is made to position the rotor of the gun transmitter.

(2) Remove cover plate on right side of elevation oil gear unit.

(3) With power turned on, insert adjusting key in hole in large spiral gear so that teeth on key mesh with teeth on smaller gear (fig. 186).

(4) Loosen the three set screws in small gear and turn key.

(5) Turning key rotates transmitter rotor, energizes oil gear unit, and moves gun.

(6) When gun is in proper orientation in elevation, the gun transmitter and director transmitter are alined.

(7) Tighten three set screws firmly and remove key.

(8) If no adjusting key is available, the adjustment may still be

## SIGHTING AND FIRE CONTROL EQUIPMENT

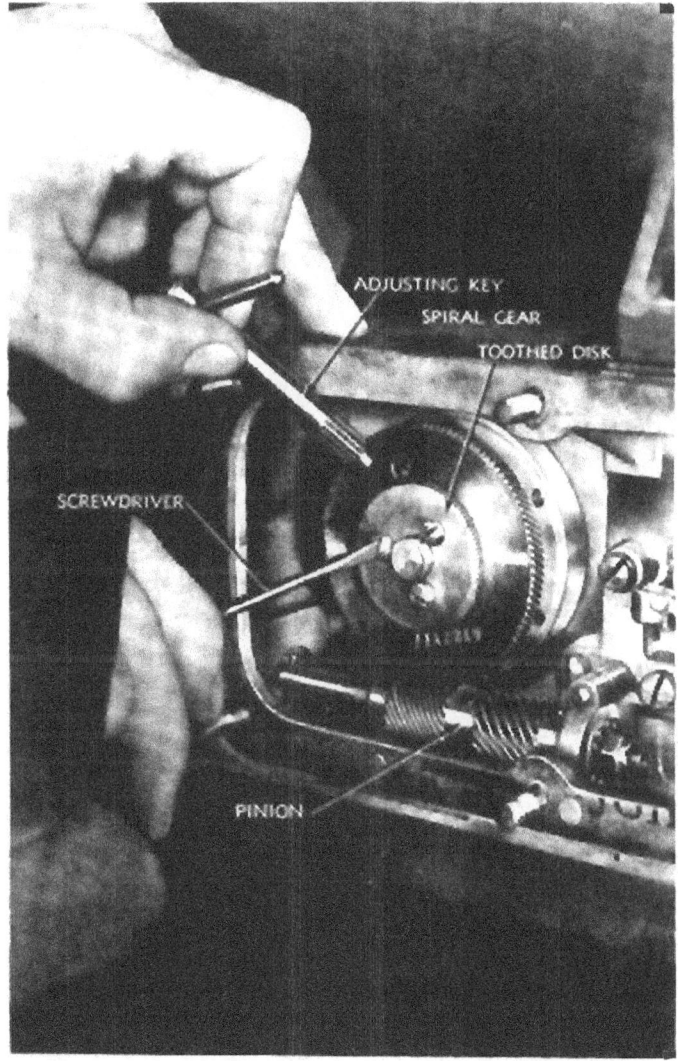

Figure 186 — Adjustment of Resetter Spiral Gear

**TM 9-252**
**98**

**40-MM AUTOMATIC GUN M1 (AA) AND 40-MM ANTIAIRCRAFT GUN CARRIAGES M2 AND M2A1**

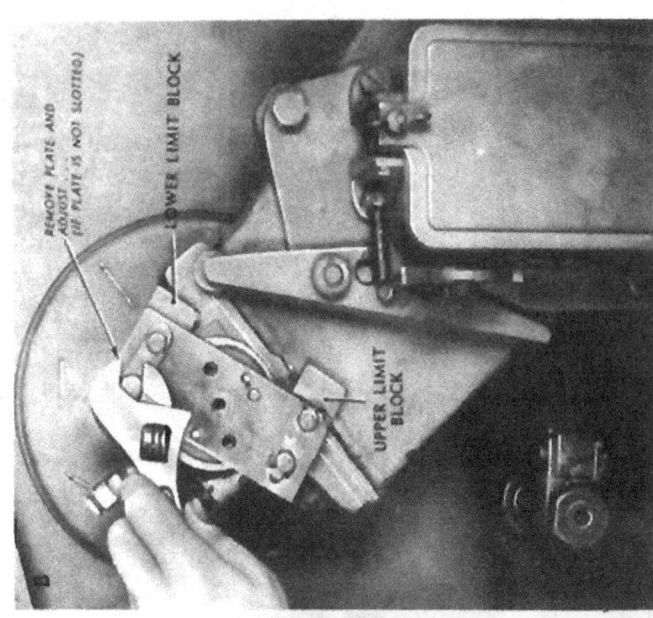

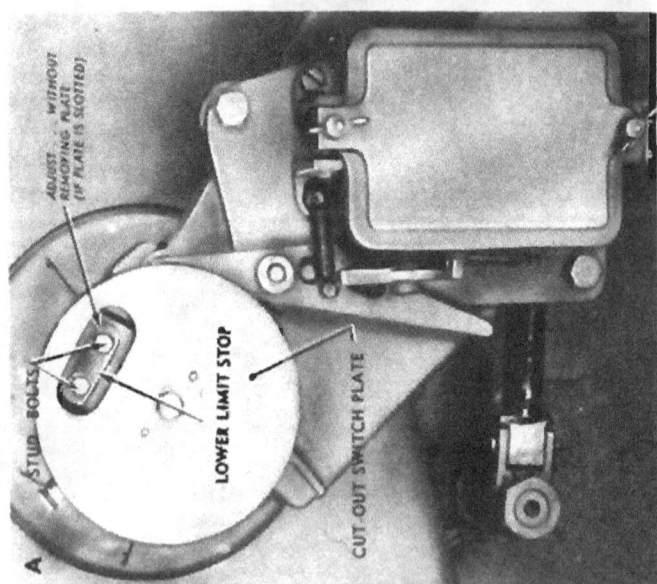

*Figure 187 — Adjusting Elevation Limit Stops*

## SIGHTING AND FIRE CONTROL EQUIPMENT

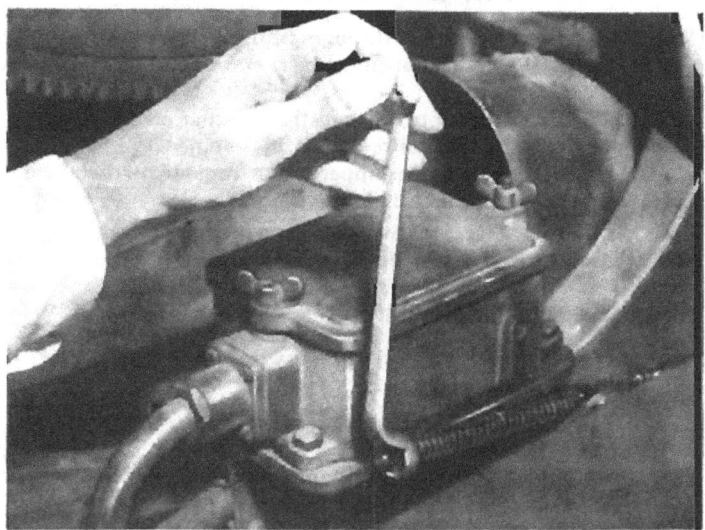

*Figure 188 — Adjusting Azimuth Switch*

made by loosening the three set screws and turning the small steel gear by hand to bring the gun into alinement. However, an accurate adjustment by hand is difficult and the key should be used if it is available.

g. **Torque Test.** This test is to check power developed by oil gear. A low rate is set in director. The gun should follow smoothly without lag. If this condition is not met, check oil level and voltage and frequency at power plant. If oil level and power are satisfactory, increase dither slightly. If unit still cannot meet the test, replacement of the oil gear unit by qualified battery specialist should be made (as per instructions contained in par. 101 b and c). The defective unit should be turned over to ordnance maintenance personnel.

h. **Adjustment of Elevation Limit Stops.**

(1) The elevation limit switch must be set so as to cut the power off the elevation oil gear before the gun hits the mechanical stops or some other obstruction such as the parapet. Normally the upper cut-off point should be 85 degrees $\pm$ 1 degree and the lower one may be anywhere between $-5$ and $+8$ degrees depending on local emplacement.

(2) To check the setting of the stops, elevate or depress the gun until the motor cuts off and then note the reading on the elevation indicator of the gun.

(3) To adjust the lower limit, loosen the two upper stud bolts.

TM 9-252
98-99

## 40-MM AUTOMATIC GUN M1 (AA) AND 40-MM ANTIAIRCRAFT GUN CARRIAGES M2 AND M2A1

These are usually accessible through a slot of the cut-out switch plate (A, fig. 187). If not, the plate must be removed and the bolts loosened (B, fig. 187).

(4) To adjust the stops, change the position of the blocks on the rotating plate until the desired limit is obtained.

(5) If necessary to adjust upper limits, remove plate, loosen lower bolts, and change position of lower blocks.

i. *Adjusting Azimuth Switch.* The switch should snap on just after the oil gear clutch is engaged by throwing the power synchronizing ("SLEWING") handle. Adjustment is made by changing the spring tension on the azimuth oil gear clutch lever (fig. 188).

## 99. MALFUNCTIONS AND CORRECTIONS, REMOTE CONTROL SYSTEM M5.

a. *General.* The operations described in paragraphs 99, 100, and 101 can be performed by the using troops under supervision of competent personnel. Any further maintenance operations, such as disassembly of oil gears or azimuth indicator, should be done only by ordnance maintenance personnel. Periodic examination of the various units should be made to insure that the system will not become inoperative due to the need of some minor adjustment or repair that could have been readily performed prior to the time of operation.

b. *If the Induction Motor in Either Oil Gear Does Not Run When Power Is Applied.* The trouble may be due to blown fuses in the elevation limit switch or azimuth switch. There are three fuses in each switch box. The fuses are accessible for replacement on removal of the switch box cover. Fuses are rated 15 amperes, 250 volts. Remove power from the system before attempting to replace fuses. Secure switch box cover tightly after fuse replacement.

c. *If No Power Is Available at Oil Gear or Director.* The trouble may be in main switch of generator. The switch should be operated slowly but firmly since, if it is flipped on, it may cause the contacts to bounce.

d. *Wiring Connections.* When checking wiring connections after repair and replacement or when searching for bad or wrong wiring connections, use the wiring diagram, described in paragraph 101 b (6) (fig. 195).

e. *Single Phase Operation of Oil Gears.*

(1) Each oil gear is driven by a 3-phase induction motor supplied with electric current through three wires from the generator. If, for any reason, one of these wires is broken so that the electric current is carried through only the other two wires, the motor will receive what is technically known as "single phase current." Under such a condition, the motor will not start. Instead, the motor merely emits

## SIGHTING AND FIRE CONTROL EQUIPMENT

a moderately loud hum without even giving a jerk or any other indication of starting. The motor will draw a heavy current of about 18 or 20 amperes from the generator which may be observed on the generator ammeter (except that if line "C" is the open one, the ammeter will read "0"). The heavy current drawn will cause the generator to slow down and probably start to "hunt." The voltage indicated on the generating unit voltmeter will have dropped to about 60 to 70 volts and the maximum adjustment of the rheostat generally will not bring the voltage back to 125 volts.

(2) Single phase operation of the motors is easily detected when the symptoms are known. When this condition does occur, the gun switch should be turned off immediately because the current drawn, if left on for more than a few seconds, is large enough to burn out the gun fuses or possibly damage the motor and generator.

(3) The most common cause for single phase operation of the motor is the failure of the gun switch to close properly. This may be overcome as follows:

*(a)* Snap the switch on quite hard.

*(b)* Open the switch box and bend the contact fingers slightly so that a smooth, easy contact is made.

*(c)* Remove burs. Burs may be removed from the fingers with sandpaper, or CLOTH, abrasive, aluminum-oxide.

NOTE: Oil or grease should not be placed on the fingers, as it interferes with good electrical contact.

(4) Another cause of single phase operation of a motor is a burned out fuse on either the gun or generator.

(5) In addition, if the generator cable plug at either gun or generator end is improperly connected so that one of the small prongs goes in the large hole, an open circuit may arise and result in single phase operation. The large prong marked "A" should go in the large hole marked "A" (fig. 170).

f. **Oil Gear Motors Running Backwards.**

(1) Reversed oil gear motors can easily be detected by the characteristic dry sound of the oil gear; by the fact that the gun gives absolutely no indication of moving; and by removing the top plug of the chain case and noting whether or not the chain is running in the direction of the arrow on the motor housing.

(2) If an oil gear motor is found to be running backwards, first check the other oil gear motor and the director motors.

(3) If all motors are running backwards, this should be corrected at the generator by interchanging any two of the three conductors attached to one end of the fuses.

(4) If only one oil gear motor is running backwards, reverse it by interchanging any two of the three leads connected to the terminals at the top of the motor (fig. 192).

g. **Reversed Tracking of the Gun.** If the gun tracks smoothly

## 40-MM AUTOMATIC GUN M1 (AA) AND 40-MM ANTIAIRCRAFT GUN CARRIAGES M2 AND M2A1

and accurately but in a direction opposite to that of the director, two of the wires to the electrical differential are interchanged. A quick correction can be made by interchanging any two of the three "S" leads or any two of the three "R" leads to the differential; however, as soon as time permits, the wiring should be checked for the original mistake and connected according to the wiring diagram (fig. 195).

h. *Basic Trouble Shooting.* The procedure for the using personnel in case of malfunctioning of any part of the remote control system is to attempt to locate the major unit in which the fault originates. Faults which appear as malfunctions of the remote control system may be in the director or in another unit of the system. To isolate the trouble to one of the three basic parts of the system, i.e., the mechanical part of the director, the electrical signalling system, or the hydraulic part of the oil gear, proceed as follows:

(1) WATCH THE DIALS ON THE DIRECTOR. The gun should follow the dials exactly. If the dials behave in the normal fashion, and the gun is erratic, the trouble is in either the hydraulic part of the oil gear or the electrical signaling system. If the dials are erratic, the trouble is in the mechanical part of the director.

(2) OPERATE THE OIL GEAR PILOT VALVE MANUALLY. If the gun responds normally, the trouble is in the electrical signaling system. If it does not, the trouble is in the hydraulic part of the oil gear.

(3) AS A ROUGH CHECK, MOVE GUN OR DIRECTOR MANUALLY AND WATCH THE DEFLECTIONS OF THE ELECTRICAL DIFFERENTIAL TORQUE ARM. If the torque arm deflects normally, the trouble is in the hydraulic part of the oil gear. If it does not, the trouble is in the electrical signaling system.

(4) Troubles in the mechanical part of the director are covered in the director manuals.

(5) Troubles in the electrical signaling system are difficult to isolate and should be referred to ordnance maintenance personnel.

(6) Troubles in the hydraulic part of the oil gear fall into three classifications as follows:

*(a) The Gun Runs Continuously in One Direction.*

1. Operate the pilot valve manually (fig. 183).
2. If this fails to affect the action, remove the pilot valve (par. 100 a).
3. Disengage azimuth oil gear clutch by blocking (fig. 176).
4. Run oil gear for 10 minutes.
5. Replace pilot valve (par. 100 b) and check operation.
6. If the trouble has not been corrected, replace the oil gear with a spare, and request ordnance maintenance personnel to repair the faulty one.

*(b) The Gun Develops Low Top Speed.*

1. This malfunction is evidenced by a tendency to "skip phase";

## SIGHTING AND FIRE CONTROL EQUIPMENT

that is, the gun jumps from one "phase" or synchronous point to another when director is tracking too rapidly. As shown on the azimuth indicator, the pointers are outside the limit lines, and therefore the gun will not synchronize with the director.

2. Run the oil gear to allow it to warm up thoroughly.

3. Measure the top speed of the gun by riding the carriage and manually holding the sensitive valve full open.

4. The gun should make one revolution in azimuth in not more than 18 seconds in either direction, and in elevation, it should make 65 degrees in not more than 4 seconds in either direction.

5. If the speed of the gun is low, add hydraulic oil to make up any deficiency.

6. Adjust the dither (par. 98 d).

7. Adjust the neutral creep (par. 98 e).

8. Again measure the top speed and, if it is still low, replace the oil gear with a spare and call ordnance maintenance personnel to repair the faulty one.

*(c) In Tracking, the Gun Is Rough or Jerky.*

1. Add hydraulic oil to make up any deficiency.

2. Adjust the dither (par. 98 d).

3. Adjust the neutral creep (par. 98 e).

4. If the gun is still rough, replace the oil gear with a spare and call ordnance maintenance personnel to repair the faulty one.

### 100. REMOVAL AND REPLACEMENT OF PILOT VALVE.

a. Removal of Pilot Valve.

(1) Remove the hex-head plug and copper asbestos gasket located at the front right corner on the top of the top case (fig. 183).

(2) Remove the six nuts from studs holding the right side plate to the top case.

(3) Remove the plate.

(4) Loosen with a screwdriver, the four screws of the differential bakelite retaining ring, located inside the left side of the top case assembly (A, fig. 189).

(5) Free the pilot valve adjusting thumbscrew from the arm of the differential (right side of the top case) (B, fig. 189). This is accomplished by pressing against the spring lip on the synchrodifferential torque arm, and by carefully moving the thumbscrew to the right until it is free of the differential torque arm.

(6) Slide the differential back (about ½ inch) by pressing against the differential arm from the right side of the top case (B, fig. 189).

(7) Grasp the thumbscrew and carefully remove the pilot valve assembly from the oil gear by lifting out through the hex-head plug hole (C, fig. 189).

(8) Replace the hex-head plug and copper-asbestos gasket in the plug hole.

**TM 9-252**
100

**40-MM AUTOMATIC GUN M1 (AA) AND 40-MM ANTIAIRCRAFT GUN CARRIAGES M2 AND M2A1**

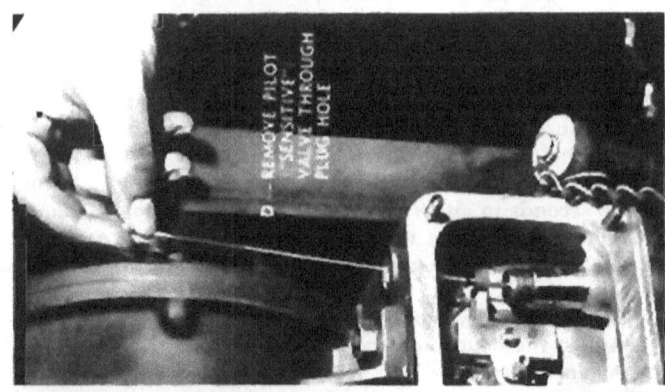
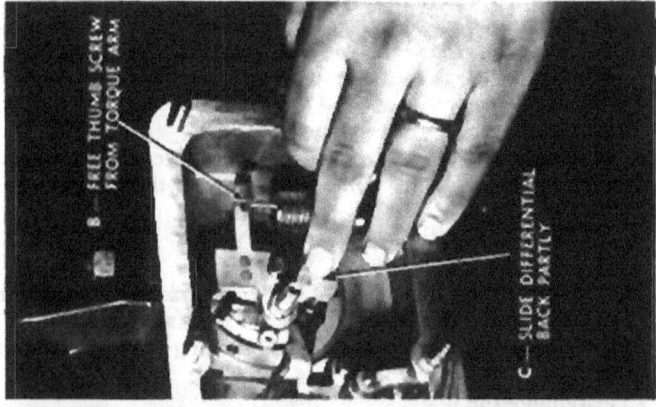
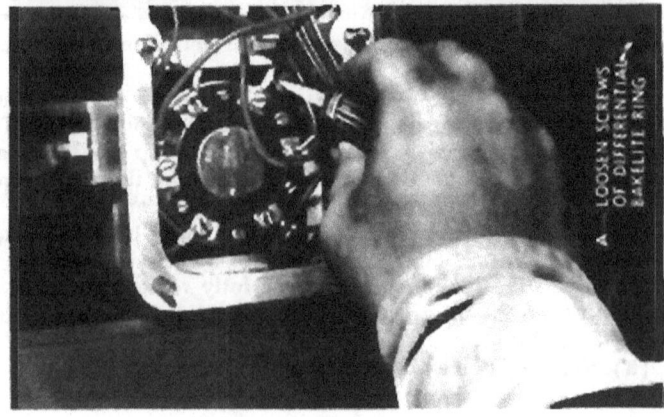

*Figure 189 — Removing the Pilot Valve*

## SIGHTING AND FIRE CONTROL EQUIPMENT

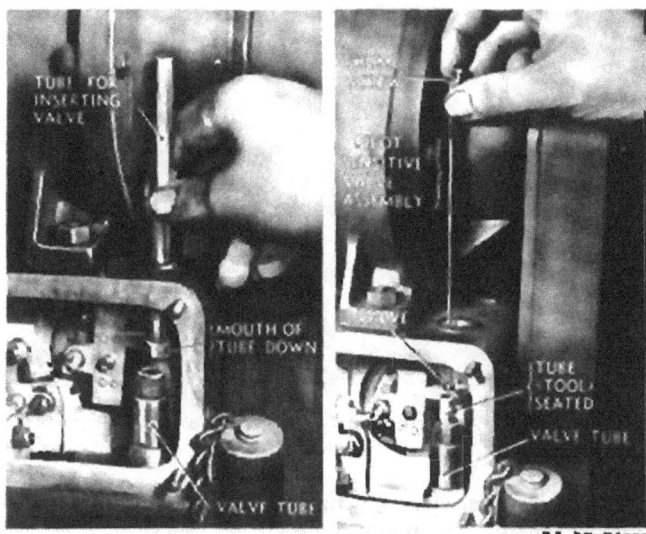

*Figure 190 — Replacing Pilot Valve*

**CAUTION:** Exercise care in handling the pilot valve. Do not touch the valve proper with the hands, as dirt and perspiration are detrimental to it. While the valve assembly is out of the oil gear wrap it in a piece of clean paper and keep it in a safe place so as to prevent burring or scratching.

b. **Replacement of Pilot Valve.**

(1) There is provided in each set of oil gear tools a tube for quickly and accurately inserting the pilot valve into the plunger of the relay valve of the oil gear. It should always be used in inserting this valve, as otherwise the valve may become caught in the mechanism and require disassembly of the oil gear to free it.

(2) Precaution. Be sure tool is absolutely clean before using. Clean with **SOLVENT**, dry-cleaning.

(3) Be sure that the differential has been pushed back as described in subparagraph a (5), above, before using the tool.

(4) Insert the tool bell-mouth down, through the ½-inch hole in the top of the transmitter and differential housing and through the sensitive valve tube until it seats over the plunger in the relay valve (A, fig. 190).

(5) Now install the sensitive valve by slipping it down into the tube (B, fig. 190).

(6) Withdraw the tube.

(7) Slide differential forward, position properly with the pin

TM 9-252
100-101

**40-MM AUTOMATIC GUN M1 (AA) AND 40-MM ANTIAIRCRAFT GUN CARRIAGES M2 AND M2A1**

*Figure 191 — Removing Top Cover from Oil Gear Motor Terminal Block*

against the bracket stop, and tighten the four screws in the bakelite retaining ring, on the opposite side (A, fig. 189).

(8) Press the valve stem back into the spring clip of the differential torque arm.

(9) Adjust the dither (par. 98 d).

(10) Make neutral ("creep") adjustment (par. 98 e).

(11) Replace cover plates.

## 101. REMOVAL, REPLACEMENT, AND CONVERSION OF OIL GEAR UNITS.

a. The using troops are permitted to perform unit replacements of oil gears in case of failure of a unit, but proper facilities must be available to prevent entrance of dust into the interior when cover plates are removed.

b. Removal of Oil Gear Units.

(1) Remove the terminal cover plate on top of the motor (fig. 191).

(2) Note the lead markings (mark the leads if there are no lead markings) to insure that the leads can be reconnected to the corresponding terminals.

(3) Loosen the three terminal screws (fig. 192) and disconnect the wires.

(4) Remove the four screws securing the cable adapter to the motor housing and withdraw the conduit from the terminal well (fig. 193).

(5) Remove the terminal side cover (fig. 194) over the transmitter and electrical differential assembly, exposing the terminal screws. This cover is on the same side as the clutch lever.

## SIGHTING AND FIRE CONTROL EQUIPMENT

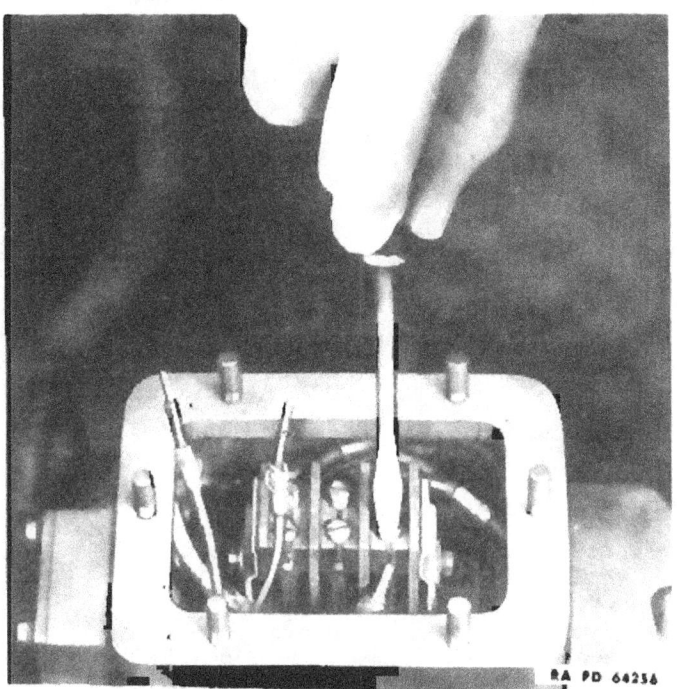

Figure 192 — Disconnecting Wires From Terminal Block

Figure 193 — Removing Conduit From Terminal Well

TM 9-252
101

**40-MM AUTOMATIC GUN M1 (AA) AND 40-MM ANTIAIRCRAFT GUN CARRIAGES M2 AND M2A1**

*Figure 194 — Removing Oil Gear Side Cover Plate*

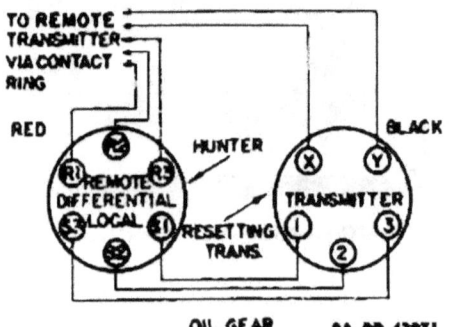

*Figure 195 — Wiring Diagram for Oil Gears*

(6) Now refer to figure 195, the wiring diagram for oil gears. Disconnect three leads (R1, R2, and R3) to the differential and two leads (X and Y) to the resetter transmitter by loosening the terminal screws. Do not disconnect the interconnecting leads (S1, S2, S3, and 1, 2, 3). Note the lead markings (mark the leads if there are no lead markings) to insure that the leads can be reconnected to their corresponding terminals.

(7) Remove the three screws securing the cable adapter to the housing (fig. 196) and withdraw the wires from the housing.

(8) Remove the four bolts securing oil gear to carriage (A, fig. 197) and the single bolt at rear of induction motor (B, fig. 197).

(9) Remove the oil gear by pulling it forward from the front of

240

## SIGHTING AND FIRE CONTROL EQUIPMENT

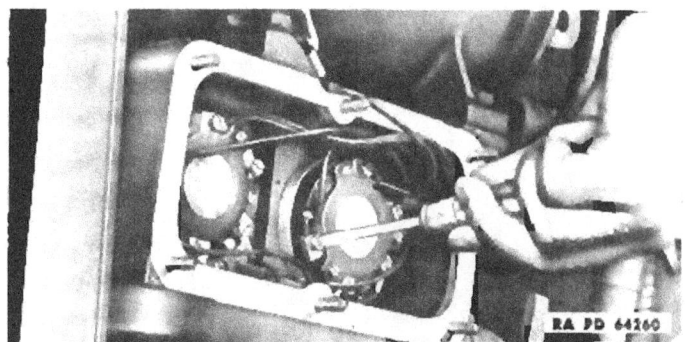

*Figure 196 — Disconnecting Wires From Oil Gear Synchro Transmitters*

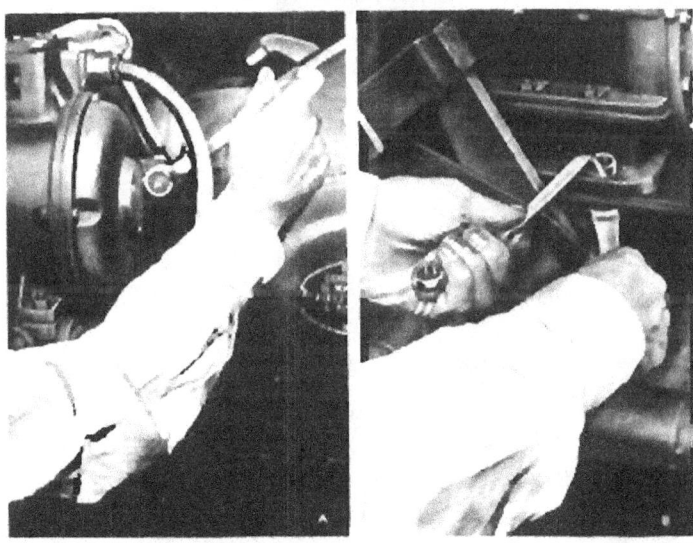

*Figure 197 — Disconnecting Oil Gear From Carriage*

the carriage, being careful not to damage the drive shaft or coupling mechanism (fig. 198).

(10) When removing an azimuth oil gear, disconnect the rod from the lower end of the clutch shaft lever.

(11) If the coupling is frozen to the spline shaft of the gun drive mechanism, then remove rubber seal from oil gear coupling by using small screwdriver (fig. 199).

TM 9-252
101

**40-MM AUTOMATIC GUN M1 (AA) AND 40-MM ANTIAIRCRAFT GUN CARRIAGES M2 AND M2A1**

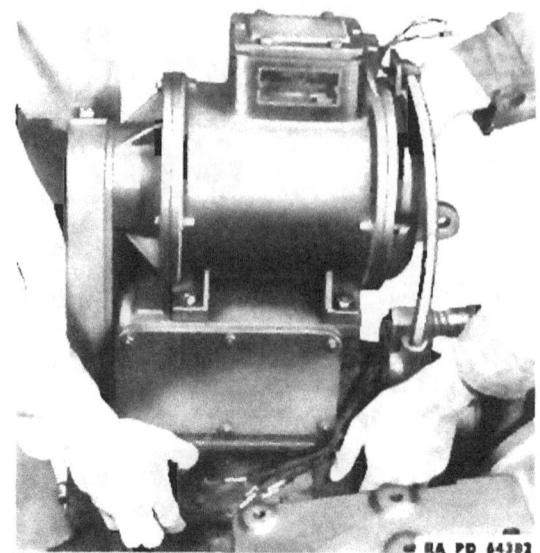

Figure 198 — Removing Oil Gear Assembly

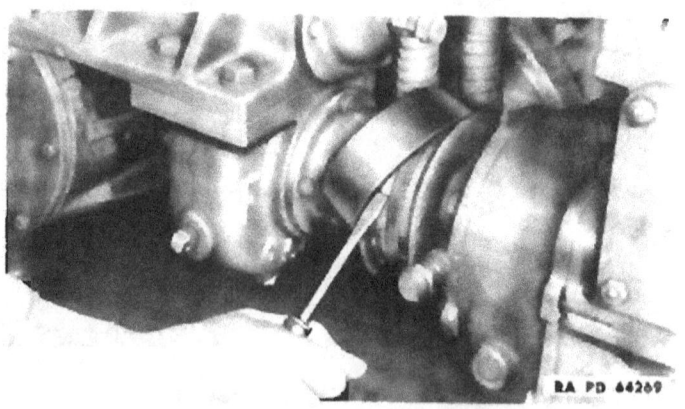

Figure 199 — Removing Rubber Seal From Oil Gear Coupling

(12) Note that in replacing a unit, a distinction must be made between the azimuth oil gear and the elevation oil gear. Although both units are similar in outward appearance, they cannot be interchanged, as their gearing ratios are different. The azimuth unit has

242

## SIGHTING AND FIRE CONTROL EQUIPMENT

a coarse (64-tooth) spiral gear on the resetter transmitter, while the elevation unit has a fine (108-tooth) spiral gear (fig. 183). In the absence of accurate name plate data, the unit can be positively identified by removing the right side (adjusting side) cover and note either the piece marks or whether the coarse or fine toothed pinion is meshed with bronze spiral gear of the gun transmitter, according to the following data: The fine is the elevation, the coarse is the azimuth one. If piece marks are visible this is the best means of identification; otherwise, count the teeth in the spiral gear.

| Type of Unit | Piece Marks | Spiral Gear |
|---|---|---|
| Elevation | A182177 | 108 teeth |
| Azimuth | A182178 | 64 teeth |

CAUTION: Oil gears must be kept level at all times in order to prevent oil from entering the electrical units above the oil gear motors.

NOTE: Oil gears may be shimmed at the bases. Care should be taken that these shims are not lost. The oil gear side plate covers and terminal block covers should be replaced to prevent dirt from entering the oil gear assembly.

c. Replacement of Oil Gear Units.
(1) Place oil gear on carriage.
(2) Insert wires into housing through proper openings.
(3) Connect leads R1, R2, and R3 to the differential and leads X and Y to the gun transmitter. Leads should have been tagged when the unit was disassembled.
(4) Connect motor leads to proper terminals in terminal well.
NOTE: The black leads from the terminal strip into the motor are for 60-cycle operation.
(5) Slide gear into position to engage the drive shaft. Be careful not to damage the drive shaft or coupling mechanism. Be sure to replace the rubber seal if it has been removed from the oil gear coupling, pulling seal over coupling towards the oil gear.
(6) Insert the five bolts that fasten the oil gear to the carriage. Start the nuts but do not tighten them. (Refer to fig. 197)
(7) The oil gear must be fastened on the carriage so that the coupling faces are parallel. For lateral alinement, shift the oil gear sidewise within the play around the bolts until the "GO" part of the coupling gage fits equally over *both* sides of the coupling (A, fig. 200) and the "NO GO" part does not (B, fig. 200). For vertical alinement, try the coupling gage on the top and bottom of the coupling. If the "GO" part of the gage does not fit equally over both top and bottom of the coupling, the oil gear may be shimmed with shim stock, or in emergency with paper, until the coupling faces are parallel. In general, however, it is not necessary to do this as the oil gear mounting plate is properly set at the proving grounds. When the oil gear is alined, tighten the bolts and, on an azimuth unit, connect the rod

TM 9-252
101

**40-MM AUTOMATIC GUN M1 (AA) AND 40-MM ANTIAIRCRAFT GUN CARRIAGES M2 AND M2A1**

*Figure 200 — Checking Coupling Faces for Parallelism With "GO" and "NO GO" Gage*

from the power synchronizing ("SLEWING") handle to the lower end of the clutch shift lever.

(8) Try the oil gear. If the motor is reversed or the gun runs backwards, proceed as in paragraph 99.

(9) After replacing an oil gear, it is necessary to reorient the gun (par. 97 c).

d. **Conversion of Elevation Oil Gear to Azimuth Oil Gear or Vice Versa.**

(1) One spare azimuth oil gear (assembled) and one spare elevation oil gear (assembled) are furnished with each eight guns.

(2) In order to convert an elevation oil gear to an azimuth unit or vice versa, the right-hand side cover plate must be removed. To identify the respective units (if nameplate data is lacking) and to be sure that the right parts are reassembled, use the data noted in subparagraph b (12), above. If piece marks are visible, this is the best means of identification; otherwise, count the gear teeth.

(3) The three screws clamping the spiral gear between the disk and adapter on the gear assembly (step 1, fig. 201) must be withdrawn, the adapter removed (step 2), and the spiral gear removed (step 3).

## SIGHTING AND FIRE CONTROL EQUIPMENT

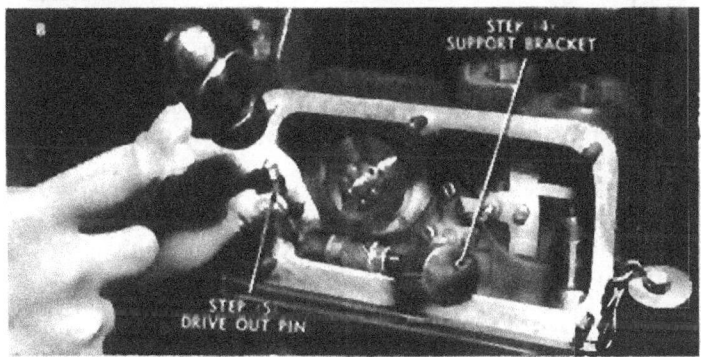

*Figure 201 — Conversion of Oil Gear from Azimuth to Elevation or Vice Versa*

TM 9-252
101

## 40-MM AUTOMATIC GUN M1 (AA) AND 40-MM ANTIAIRCRAFT GUN CARRIAGES M2 AND M2A1

(4) The taper pin securing the double pinion (fine pinion for elevation, coarse pinion for azimuth) must be driven out. Note the following precautions and instructions:

(a) Be sure to drive the taper pin through from the small end. Rotate the pinion and note carefully the relative size of the pin ends, until certain which is the smaller end.

(b) Insert a screwdriver or other firm tool under the pinion shaft (step 4, fig. 201) in order to support the shaft while driving out the pin. This will prevent bending or breaking the aluminum bearing bracket.

(c) Drive out the pin using a drift tool or a drive pin punch (step 5, fig. 201).

(5) Slide pinion over to the other side of the shaft (engaging the fine pinion if it is desired to convert to elevation, and coarse pinion if it is desired to convert to azimuth) (step 6, fig. 201).

(6) Secure in place by inserting taper pin into the hole provided (step 7, fig. 201)

(7) Mount the appropriate spiral gear (mount the fine spiral gear if converting to elevation, or the coarse spiral gear if converting to azimuth).

(8) Replace the disk.

(9) Secure the gear and disk to the adapter by means of the three screws that were withdrawn, taking care to mesh the gear with the correct pinion, as instructed in step (5), above.

(10) Check backlash between the pinion and gear. Adjust so that the pressure between teeth of the pinion and gear is as light as possible while there is no play between the gears. The adjustment is made by loosening the screws holding the *bearing bracket*, pressing the bracket upward and reclamping the screws. (To be done only by battery electrician.)

(11) Where only one lug is provided on the bell housing, it will be necessary on those units to remove the four nuts and studs holding the rear bell housing. In order to remove the studs with the nuts, *mutilate threads* on ends of studs so that nuts will not screw completely off. The studs will come out upon exerting pressure on nuts. After removing nuts with studs, rotate bell housing 180 degrees and secure in place.

(12) Precaution. Do not under any circumstances pull bell housing off motor. If this is done the rear motor bearings will fall out and replacement of the bearings necessitates use of a special tool which is not issued to the using arms. The bell housing is rotated to enable fastening the upper portion of the oil gear unit to carriage (bell housings which are equipped with two brackets need not be rotated).

## SIGHTING AND FIRE CONTROL EQUIPMENT

(13) Interchange the pipe plug and grease fitting in the clutch housing cover and secure in place.

(14) Remove the cotter pin on the shaft pin which supports the auxiliary locking lever (fig. 174), drive out the shaft pin, and remove the locking lever. Insert the spare sleeve, drive the shaft pin back in place, and secure it by means of the cotter pin.

(15) An azimuth unit can be converted to an elevation unit by reversing the above replacements and assemblies.

(16) When spare elevation oil gear is changed over into an azimuth oil gear and replaces the azimuth unit on the gun, be sure to switch nameplates. After having done this, change azimuth unit, which has just been removed from the gun, into an elevation oil gear. Nameplates will then be correct.

c. Tools and Equipment. A tool box containing tools and equipment for maintenance of the oil gear is furnished with the system. No screwdriver is provided as there is usually a screwdriver in the gun carriage tool box. The following tools are supplied. When not in use, these tools should always be kept in the tool box provided.

Coupling gage (for assembling oil gear to carriage)
Oil gun (for adding oil to unit)
Adjusting key (for adjusting resetter transmitter)
Spanner (for changing filter bobbin assembly)
Filling tube (for inserting sensitive valve)
12-point box wrench ($\frac{3}{8}$ and $\frac{7}{16}$)
12-point box wrench ($\frac{5}{8}$ and $\frac{3}{4}$)

## 102. LUBRICATION OF REMOTE CONTROL SYSTEM M5.

a. Lubrication and Replenishing of Oil.

(1) Cleanliness is of the utmost importance in handling the oil for the oil gears. When replenishing the oil supply of the oil gears, the oil must be taken from a container whose seal is unbroken. Dust or moisture must not be allowed to get into the oil while refilling, and the remaining oil in the container should be used for some other purpose or thrown away. The oil gun must be kept clean, and should be used only for the prescribed oil (step (3), below).

(2) It is a good practice to have one man replenish the oil in all oil gears in the battery at least once a week by carrying the can of oil from gun to gun and thereby minimize the waste of oil to a part of a can per battery, instead of a part of a can per gun.

(3) Since all oil gears leak oil around the pump and motor shafts, it is necessary to replenish the oil in the main case by pumping OIL, hydraulic, U. S. Army Specification 2-79A, through the filter until oil flows from the overflow hole in the side of the filling tube. This should be done once a week (subpar. b below). As some oil gears become air locked, it may be necessary to loosen the pilot valve filling plug to let the air escape from the case while the oil is pumped in.

TM 9-252
102

**40-MM AUTOMATIC GUN M1 (AA) AND 40-MM ANTIAIRCRAFT GUN CARRIAGES M2 AND M2A1**

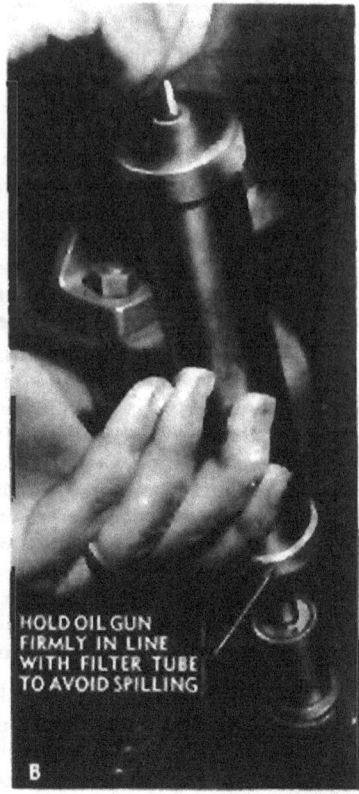

*Figure 202 — Pumping Oil Through the Filter of Oil Gear*

(4) The oil gear, when filled with oil, must always be kept upright; otherwise, oil will get into the electrical control elements and cause damage. If, through accident, the oil does get into the electrical control compartment, remove the side covers from this compartment and wipe off the oil.

(5) Precaution. Unless the proper procedure is followed, the mechanic filling the oil gear may erroneously decide that the gear is filled, whereas the oil which appears at the location of the tell-tale hole in the oil filter assembly may actually have been spilled in the filling operation. If the oil gun is not centered in the bushing in the oil filter, spilling will occur. The oil gun can be centered in this bushing by first elevating the 40-mm gun barrel to its maximum position. When filling the elevation oil gear (B, fig. 202), hold oil gun firmly in line with the filter tube.

248

## SIGHTING AND FIRE CONTROL EQUIPMENT

b. Changing Oil.
(1) Remove pilot valve (par. 100 a).
(2) Leave elevation oil gear disengaged when changing oil in elevation oil gear.
(3) Block azimuth oil gear clutch (fig. 178) in disengaged position when changing oil in azimuth oil gear.
(4) Turn on oil gear and allow it to run for 10 minutes.
(5) Stop oil gear.
(6) Immediately remove drain plug under filling tube.
(7) Allow oil to drain from main case.
(8) Replace drain plug.
(9) Tighten securely.
(10) Replace pilot valve (par. 100 b).
(11) Refill by pumping OIL, hydraulic, through filter (fig. 202).
(12) Stop when oil flows from overflow hole in the side of the filling tube.
(13) Do not flush oil gear with solvent or gasoline as this would dilute the oil.
(14) Make neutral (creep) adjustment (par. 98 e).
(15) Reorient the gun (par. 97 c).

c. Renewing Filter Bobbin Assembly. The filter (filling tube) contains a filter bobbin assembly which may require replacement after several oil changes. The filter bobbin assembly consists of a bobbin shaped metal tube with a muslin strip wrapped around a perforated portion to form the filter element. To replace the filter bobbin assembly, proceed as follows:
(1) Remove cover (fig. 202).
(2) Using the spanner tool provided (A, fig. 203), unscrew the retainer at the top of the filter opening filling tube (B, fig. 203).
(3) Lift out the old filter bobbin assembly by grasping the top with the fingers and pulling out (C, fig. 203).
(4) Wash in SOLVENT, dry-cleaning. If the filter is damaged and ineffective or too dirty to be cleaned satisfactorily, replace it with a new one.
(5) Screw the retainer back into place.
CAUTION: Care must be taken during the foregoing operations to prevent entrance of dirt into the interior parts.

d. Chain Case and Clutch Housing.
(1) Drain oil from chain case daily, by removing level plug in chain case.
(2) Drain oil from clutch housing weekly, by removing level plug in clutch housing.
(3) Do not remove chain case cover. (Chains are not to be greased.)
(4) In replacing plugs, use care not to tighten sufficiently to strip threads in the aluminum chain case.

**TM 9-252**
**102**

**40-MM AUTOMATIC GUN M1 (AA) AND 40-MM ANTIAIRCRAFT GUN CARRIAGES M2 AND M2A1**

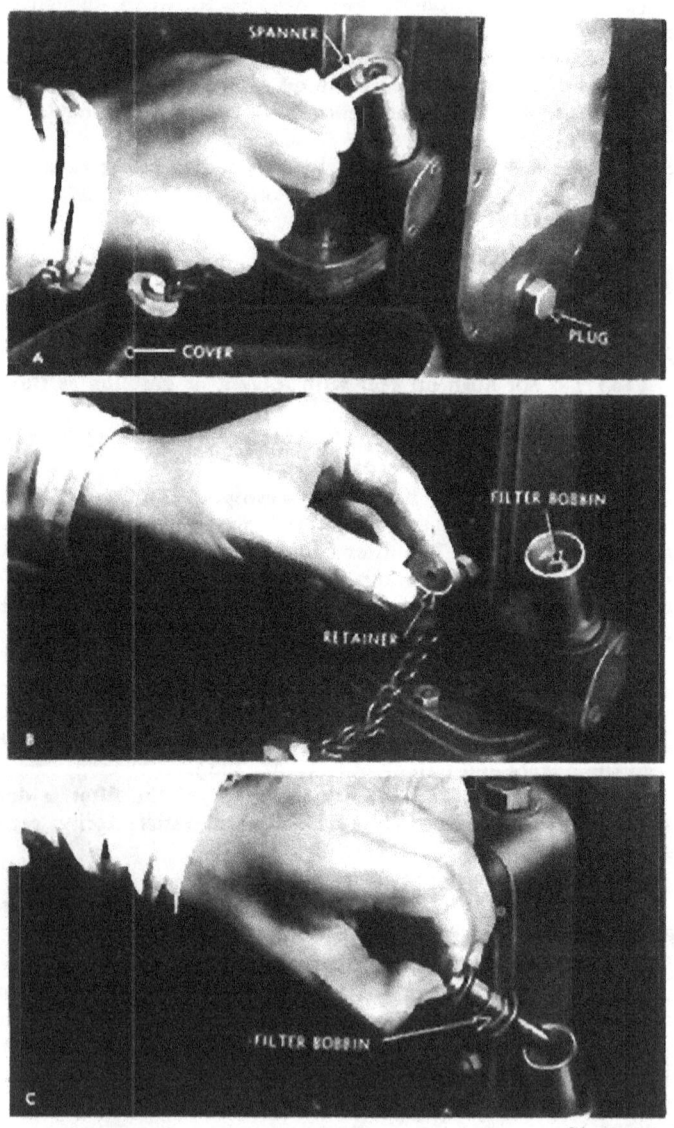

*Figure 203 — Renewing Filter Bobbin Assembly*

## SIGHTING AND FIRE CONTROL EQUIPMENT

*Figure 204 — Lubrication of Transmitter Spiral Gear*

e. **Lubrication of Transmitter Spiral Gear.**

(1) The resetter transmitter spiral gear (fig. 204) is lubricated by a cork lubricating ball located in a recess under the driving pinion.

(2) Fill this recess with hydraulic oil as prescribed in lubrication section.

(3) The cover over the spiral gear must be removed to permit lubrication (fig. 204).

(4) When lubricating spiral gear, note condition of cork ball. The ball must be replaced if it becomes "chewed up" (to be done by ordnance maintenance personnel).

(5) Sufficient oil to float the cork ball up to the opening must be kept in the oil recess.

f. **Azimuth Indicator.**

(1) The azimuth indicator has an oil cup fitted with a felt pad, located in top of housing.

(2) Every 6 months, apply 6 to 8 drops of OIL, lubricating, for aircraft instruments and machine guns, to these oil cups. Oil very sparingly, as excess oil will damage the indicator.

g. **Contact Ring.** Keep the area around the contact ring clean from oil, water, and dirt. In order to perform the cleaning operation, a portion of the cartridge case chute has to be removed.

# 40-MM AUTOMATIC GUN M1 (AA) AND 40-MM ANTIAIRCRAFT GUN CARRIAGES M2 AND M2A1

## Section XI

## AMMUNITION

### 103. GENERAL.

a. Ammunition for the GUN, automatic, 40-mm, M1, is issued in the form of fuzed complete rounds of fixed ammunition. The term "fixed" signifies that the propelling charge is not adjustable and that the round is loaded into the gun as a unit. The cartridge case, which contains the propelling charge and primer, is crimped rigidly to the projectile. A complete round includes all of the ammunition components (cartridge case, primer, propelling charge, and projectile) required to fire the gun once. All high-explosive projectiles are provided with point-detonating fuzes. In firing, the cartridge case is extracted and ejected automatically.

### 104. NOMENCLATURE.

a. Standard nomenclature is used herein in all references to specific items of issue. Its use for all purposes of record is mandatory.

### 105. FIRING TABLES.

a. For available firing tables, see section XV.

### 106. CLASSIFICATION.

a. Dependent upon the type of projectile, ammunition for this gun is classified as high-explosive (HE), high-explosive-tracer (HE-T), armor-piercing-tracer (AP-T), target-practice-tracer (TP-T), or drill. The projectile of the high-explosive round is relatively thin-walled and contains a high-explosive filler for fragmentation and blast effect. The projectile of the armor-piercing round is a solid shot, for penetration of armor-plate or other highly resistant targets. The practice ammunition has a service propelling charge but the projectile is inert and fitted with a dummy or inert fuze. It is provided for training in markmanship. All types manufactured for Army use have tracer elements in the projectile base to facilitate observation of the path of flight. The tracer assembly of the high-explosive shell is known as a shell-destroying tracer because, in addition to providing a visible trace for observation, it will detonate the bursting charge at the end of the trace burnout should the projectile fail to strike the target or fail to explode as a result of impact. The drill cartridge is a completely inert assembly.

### 107. IDENTIFICATION.

a. General. Ammunition is identified by painting and marking which appears on all original packing containers and, when practi-

## AMMUNITION

cable, on the items themselves. Identifying markings on packing boxes are shown in figure 208, and are listed with other markings for shipment in paragraph 113. It will be noted that the marking includes the muzzle velocity. When removed from packing, the ammunition is identified as indicated in the paragraphs below when of Army procurement. In other cases, painting and marking may differ in form but essentially the same information will be given.

b. Mark or Model. To identify a particular design, a model designation is assigned at the time the model is classified as an adopted type. This model designation is included in the marking of the item. Components originated by the Army are designated by the letter "M" followed by an Arabic numeral, for example, "M64." Modifications are signified by adding the letter "A" and appropriate Arabic numeral. Thus, "M64A1" signifies the first modification of an item for which the original designation was "M64." Components originated by other services are designated in accordance with the practices of those services. (Navy practice is to use the word "Marks", abbreviated "Mk.," followed by either a Roman or an Arabic numeral.)

c. Ammunition Lot Number. In addition to the lot number assigned to every ammunition component and usually stamped thereon, assembled complete rounds are assigned an *ammunition lot number* at the time of assembly. This ammunition lot number is stenciled on all original packing containers and, when practicable, on the assembly itself. It is required for all purposes of record, including reports on condition, functioning, and accidents, in which the ammunition is involved. For the most uniform results in firing, successive rounds should be from the same ammunition lot.

d. Marking. The 40-mm rounds are marked for identification as follows:

(1) STENCILED ON THE PROJECTILE.
  (a) Caliber and type of cannon in which fired.
  (b) Type and model of projectile.
  (c) Kind of filler (on HE projectiles).
  (d) "WITH TRACER" (when applicable). This marking will include the model of tracer.
  (e) Ammunition lot number and loader's initials.‡
  (f) Lot number of filled projectile. (Ordinarily the projectile lot number is not required after the complete round is assembled. Hence, it is stenciled below the rotating band, in which position it is covered by the neck of the cartridge case.)

(2) STAMPED ON THE ROTATING BAND.
  (a) Lot number of empty projectile.

---
‡ — Ammunition lot number formerly appeared on the base of the cartridge case.

TM 9-252
107-108

**40-MM AUTOMATIC GUN M1 (AA) AND 40-MM ANTIAIRCRAFT GUN CARRIAGES M2 AND M2A1**

 (b) Year of manufacture.
 (c) Initials or symbol of manufacturer.
 (d) Caliber and model of shot or shell.
 (3) ON THE BASE OF THE CARTRIDGE CASE.
 (a) Caliber and model of cartridge case (stamped).
 (b) Cartridge case lot number, manufacturer's initials and year manufactured (stamped).
 (4) ON THE FUZE (STAMPED IN BODY).
 (a) Model designation of fuze.
 (b) Manufacturer's initials.
 (c) Lot number of fuze.
 (d) Year of manufacture.

 *e.* **Painting.** Projectiles are painted primarily to prevent rust; the secondary purpose is to provide, by the color, a ready means of identification as to type. The color scheme for 40-mm ammunition is as follows:

| | |
|---|---|
| High-explosive | Olive-drab, marking in yellow. |
| Armor-piercing (w/o explosive) | Black, marking in white. |
| Target-practice | Blue, marking in white. |
| Drill (dummy) | Black, except bronze parts; marking in white. |

## 108. CARE, HANDLING, AND PRESERVATION.

 *a.* **General.** Ammunition is packed to withstand conditions ordinarily encountered in the field. Care must be observed to keep packing containers from becoming broken or damaged. All broken containers must be repaired immediately and careful attention given to the transfer of all markings to the new parts of the container.

 *b.* Since explosives are adversely affected by moisture and high temperature, due consideration should be given the following:
 (1) Do not break the moisture-resistant seal until ammunition is to be used. Ammunition removed from airtight containers, particularly in damp climates, is apt to corrode, thereby causing the ammunition to become unserviceable.
 (2) Protect the ammunition, particularly fuzes, from sources of high temperature, including the direct rays of the sun. More uniform firing is obtained if the rounds are at the same temperature.

 *c.* Explosive ammunition must be handled with appropriate care at all times. The explosive elements in primers and fuzes are particularly sensitive to undue shock and high temperature.

 *d.* Ammunition should be protected from mud, sand, dirt, and water. If the rounds become wet or dirty, they should be wiped off at once. Verdigris or light corrosion should be wiped off. Am-

## AMMUNITION

munition should not be polished, however, to make it look better or brighter.

*e.* Rounds prepared for firing but not fired will be returned to their original condition and packing, and appropriately marked. Such ammunition will be used first in subsequent firing in order that stocks of opened packings may be kept at a minimum.

*f.* Do not handle duds. Because their fuzes are armed, and hence extremely dangerous, duds will not be moved or turned, but will be destroyed in place in accordance with TM 9-1900.

### 109. AUTHORIZED ROUNDS.

*a.* Ammunition authorized for use in GUN, automatic, 40-mm, M1, is listed in table I and illustrated in figures 205 to 207. It will be noted that the nomenclature (standard nomenclature) completely identifies the round. The numbers in parentheses preceding the nomenclature refer to assembly numbers in table II, which contains ammunition data.

**TM 9-252**
**109**
**40-MM AUTOMATIC GUN M1 (AA) AND 40-MM ANTIAIRCRAFT GUN CARRIAGES M2 AND M2A1**

TABLE I — AUTHORIZED ROUNDS

**SERVICE AMMUNITION**
(1) CARTRIDGE, AP-T, M81A1, 40-mm AA. guns
— CARTRIDGE, AP-T, M81A1, steel case, 40-mm AA. guns
(2) CARTRIDGE, AP-T, M81, 40-mm AA. guns
(3) CARTRIDGE, HE, Mk. I (Navy), w/FUZE, P.D., Mk. 27 (Navy), 40-mm AA. guns
(4) CARTRIDGE, HE-T (SD, M3), Mk. II, w/FUZE, P.D., Mk. 27 (Navy), 40-mm AA. guns (muzzle velocity 2,700 f/s)
— CARTRIDGE, HE-T (SD, M3), Mk. II, steel case, w/FUZE, P.D., Mk. 27 (Navy), 40-mm AA. guns (muzzle velocity 2,700 f/s)
(5) CARTRIDGE, HE-T (SD), Mk. II, w/FUZE, P.D., M71, 40-mm AA. guns (muzzle velocity 2,700 f/s)
(6) CARTRIDGE, HE-T (SD, M3), Mk. II, w/FUZE, P.D., M64A1, 40-mm AA. guns (muzzle velocity 2,700 f/s)
— CARTRIDGE, HE-T (SD, M3), Mk. II, steel case, w/FUZE, P.D., M64A1, 40-mm AA. guns (muzzle velocity 2,700 f/s)
(7) CARTRIDGE, HE-T (SD, No. 12), Mk. II, w/FUZE, P.D., Mk. 27 (Navy), 40-mm AA. guns (muzzle velocity 2,700 f/s)
— CARTRIDGE, HE-T (SD, No. 12), Mk. II, steel case, w/FUZE, P.D., Mk. 27 (Navy), 40-mm AA. guns (muzzle velocity 2,700 f/s)
(8) CARTRIDGE, HE-T (SD, M3), Mk. II, w/FUZE, P.D., M64A1, 40-mm AA. guns (muzzle velocity 2,870 f/s)
— CARTRIDGE, HE-T (SD, M3), Mk. II, steel case, w/FUZE, P.D., M64A1, 40-mm AA. guns (muzzle velocity 2,870 f/s)
(9) CARTRIDGE, HE-T (SD, No. 12), Mk. II, w/FUZE, P.D., M64A1, 40-mm AA. guns (muzzle velocity 2,870 f/s)
— CARTRIDGE, HE-T (SD, No. 12), Mk. II, steel case, w/FUZE, P.D., M64A1, 40-mm AA. guns (muzzle velocity 2,870 f/s)
(10) CARTRIDGE, HE-T (SD, No. 12), Mk. II, w/FUZE, P.D., Mk. 27 (Navy), 40-mm AA. guns (muzzle velocity 2,870 f/s)
— CARTRIDGE, HE-T (SD, No. 12), Mk. II, steel case, w/FUZE, P.D., Mk. 27 (Navy), 40-mm AA. guns (muzzle velocity 2,870 f/s)
(11) CARTRIDGE, HE-T (SD), Mk. II, w/FUZE, P.D., 251, Mk. I, 40-mm AA. guns (muzzle velocity 2,870 f/s)
— CARTRIDGE, HE-T (SD), Mk. II, steel case, w/FUZE, P.D., 251, Mk. I, 40-mm AA. guns (muzzle velocity 2,870 f/s)

**PRACTICE AMMUNITION**
(12) CARTRIDGE, TP-T, M91, w/FUZE, dummy or inert, M *, 40-mm AA. guns (muzzle velocity 2,700 f/s)
(13) CARTRIDGE, HE-T, Mk. II, inert loaded, w/FUZE, dummy or inert, M *, 40-mm AA. guns (muzzle velocity 2,700 f/s)

**DRILL AMMUNITION**
(14) CARTRIDGE, drill, M17, 40-mm AA. guns

\* — FUZE, dummy, M69; or inert FUZE, P.D., M64, M64A1, Mk. 27, 251 Mk. I, or other suitable types may be assembled to the projectile.

**TM 9-252**
**109**

## AMMUNITION

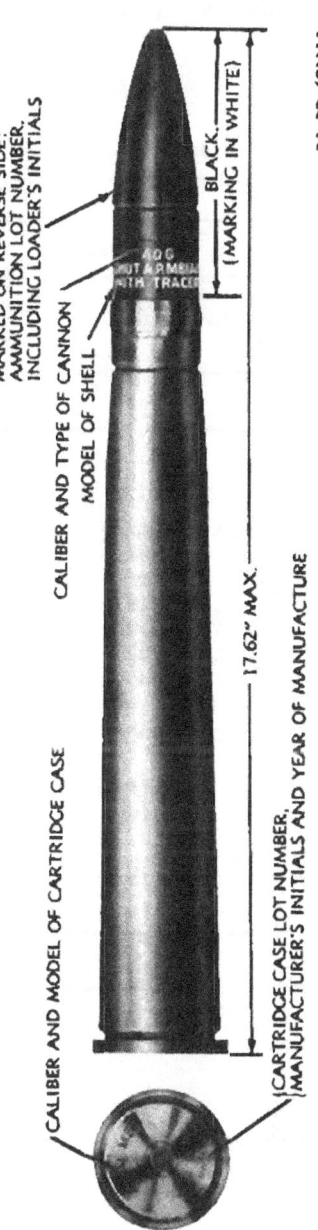

*Figure 205 — Cartridge, AP-T, M81A1, 40-mm AA Guns*

257

TM 9-252
109

## 40-MM AUTOMATIC GUN M1 (AA) AND 40-MM ANTIAIRCRAFT GUN CARRIAGES M2 AND M2A1

### TABLE II — DATA ON AMMUNITION

NOTE: Data given for the rounds listed below are based on standard assembled from similar components manufactured by other services

| | Kind | Type | PROJECTILE | | Charge | | COMPLETE ROUND | |
|---|---|---|---|---|---|---|---|---|
| | | | Model | Weight as fired (lb) | Kind | Weight (lb) | Weight (lb) | Length (in.) |

SERVICE

| | Kind | Type | Model | Weight (lb) | Kind | Weight (lb) | Weight (lb) | Length (in.) |
|---|---|---|---|---|---|---|---|---|
| 1 | CARTRIDGE | AP-T | M81A1 | 1.96 | None | — | 4.58 | 17.62 |
| 2 | CARTRIDGE | AP-T | M81 | 1.96 | None | — | 4.53 | 17.62 |
| 3 | CARTRIDGE | HE§ | Mk.I(Navy) | 2.0 | TNT | 0.1 | 4.6 | 17.60 |
| 4 | CARTRIDGE | HE-T(SD,M3) | Mk.II | 2.06 | Tetryl | 0.05 | 4.82 | 17.62 |
| 5 | CARTRIDGE | HE-T(SD) | Mk.II | 2.06 | Tetryl | 0.05 | 4.82 | 17.62 |
| 6 | CARTRIDGE | HE-T(SD,M3) | Mk.II | 1.93 | TNT | 0.14 | 4.53 | 17.62 |
| 7 | CARTRIDGE | HE-T(SD,M3) | Mk.II | 1.93 | TNT | 0.168 | 4.82 | 17.62 |
| 8 | CARTRIDGE | HE-T(SD,M3) | Mk.II | 2.06 | Tetryl | 0.05 | 4.49 | 17.62 |
| 9 | CARTRIDGE | HE-T(SD,No. 12) | Mk.II | 1.96 | TNT | 0.14 | 4.49 | 17.62 |
| 10 | CARTRIDGE | HE-T(SD,No. 12) | Mk.II | 1.93 | TNT | 0.168 | 4.78 | 17.62 |
| 11 | CARTRIDGE | HE-T(SD) | Mk.II | 1.93 | TNT | 0.15 | 4.60 | 17.61 |

PRACTICE

| 12 | CARTRIDGE | TP-T% | M91 | | None | — | | 17.62 |
| 13 | CARTRIDGE | HE-T% | Mk.II | 1.99 | Inert-Loaded | — | 4.68 | 17.60 |

DRILL

| 14 | CARTRIDGE | Drill | M17 | — | — | — | 4.53 | 17.62 |

AP-T — armor-piercing-tracer
FNH — flashless nonhygroscopic
f/s — feet per second
gr. — grain
HE-high-explosive
HE-T(SD) — high-explosive-tracer (shell destroying)
P.D. — point-detonating
perc. — percussion
TP-T — target-practice-tracer

## AMMUNITION

### FOR 40-MM AA. GUN M1

components. Complete rounds may be found which have been interchanged components will be identified by the marking thereon.

| FUZE | | *PROPELLING CHARGE | | | #PRIMER | |
|---|---|---|---|---|---|---|
| Type and Model | Action | Type of Powder | Weight (lb) | Muzzle Velocity f/s | Model | Type |
| **AMMUNITION** | | | | | | |
| None | — | FNH | 0.65 | 2870 | M38A1 | 55-gr. perc. |
| None | — | FNH | 0.65 | 2870 | M23A2 | 20-gr. perc. |
| P.D., Mk. 27 (Navy) | Superquick | FNH | 0.66 | | Mk. 22 | 64-gr. perc. |
| P.D., Mk. 27 (Navy) | Superquick | FNH | 0.72 | 2700 | M38A1 | 55-gr. perc. |
| P.D., M71 | Superquick | FNH | 0.72 | 2700 | M38A1 | 55-gr. perc. |
| P.D., M64A1 | Superquick | FNH | 0.72 | 2700 | M38A1 | 55-gr. perc. |
| P.D., Mk. 27 (Navy) | Superquick | FNH | 0.72 | 2700 | M38A1 | 55-gr. perc. |
| P.D., M64A1 | Superquick | FNH | 0.68 | 2870 | M38A1 | 55-gr. perc. |
| P.D., M64A1 | Superquick | FNH | 0.68 | 2870 | M38A1 | 55-gr. perc. |
| P.D., Mk. 27 (Navy) | Superquick | FNH | 0.68 | 2870 | M38A1 | 55-gr. perc. |
| P.D., 251, Mk. 1 | Superquick | FNH | 0.68 | 2870 | M38A1 | 55-gr. perc. |
| **AMMUNITION** | | | | | | |
| Dummy or inert, M‡ | — | FNH | 0.72 | 2700 | M38A1 | 55-gr. perc. |
| Dummy or inert, M‡ | — | FNH | 0.72 | 2700 | M38A1 | 55-gr. perc. |
| **AMMUNITION** | | | | | | |
| None | — | — | — | — | — | — |

\* — Standard cartridge case is CASE, cartridge, 40-mm, M25. Substitute is M25B1, a steel case weighing 0.26 pounds less than the standard brass case.

\# — Standard primer is PRIMER, percussion, M38A1. Earlier standards are: The M23A2 and the No. 12, Mk. II/L/ (British). Alternatives are: The M38B2 and Mk. 22 (Navy).

‡ — FUZE, dummy, M69; or inert FUZE, P.D., M64, M64A1, Mk. 27, 251 Mk. I, or other suitable types may be used.

% — Projectile has a tracer for observation purposes only.

§ — This shell has the Navy Shell-destroying Tracer Mk. 8 or Mk. 10.

## 40-MM AUTOMATIC GUN M1 (AA) AND 40-MM ANTIAIRCRAFT GUN CARRIAGES M2 AND M2A1

### 110. PREPARATION FOR FIRING.

a. Charger clips loaded with rounds are ready for insertion in the automatic feed mechanism of the gun. Rounds packed in fiber containers, when removed therefrom, need only to be loaded in the charger clips.

### 111. DESCRIPTION AND DATA.

a. Data on the ammunition is given in table II which contains ammunition data. Additional information on components is given in the paragraphs below.

b. **Armor-piercing Projectile.** SHOT, AP-T, M81A1, is a steel projectile of the monobloc type. The body is solid except for a cavity in the base which holds the tracer composition. An ogival windshield is secured directly to the body of the shot by a 360-degree crimp. SHOT, AP-T, M81, is similar to the M81A1 Shot except that the windshield is attached to an adapter which is soldered to the body of the shot.

c. **High-explosive Projectile.** The body of SHELL, HE-T (SD), Mk. II, is a hollow steel casing with a conical nose and "boat-tailed" (conical) base. The nose and the base are threaded internally, the nose to receive the point-detonating fuze, and the base to receive TRACER and IGNITER, Shell, No. 12, Mk. I/L/, internal, or TRACER, M3. The No. 12" tracer, a British design, consists of a primer, an igniting charge, a red tracer composition, and a relay igniting charge. Functioning of the tracer is initiated by set-back upon firing. The tracer burns with a visible trace for about 7 to 10 seconds. As the tracer composition burns out, the relay igniting charge is ignited, detonating the bursting charge of the shell unless prior detonation has been caused by functioning of the fuze. In the M3 Tracer, a less complex type, the primer is omitted and the powder train is made up of one igniter charge, three charges of red tracer composition, and one relay igniting charge. Functioning is initiated by the propelling charge upon firing; the tracer then burns for 12 to 14 seconds. When fitted with the M3 Tracer, the shell is loaded with tetryl. Shells with the No. 12 tracer have a TNT filler. The bursting charge is approximately the same weight for all shells but the method of loading varies slightly for each model of fuze. This is necessary to provide a suitable charge cavity for the particular fuze assembled to the shell.

d. **Cartridge Cases and Primers.** CASE, cartridge, 40-mm, M25, or M25B1, is used in 40-mm ammunition of Army design. The M25B1 case differs from the M25 in that it is made of steel instead of drawn brass, has a thinner head and primer seat, and weighs approximately 0.26 pound less. The primer used in these cases is,

## AMMUNITION

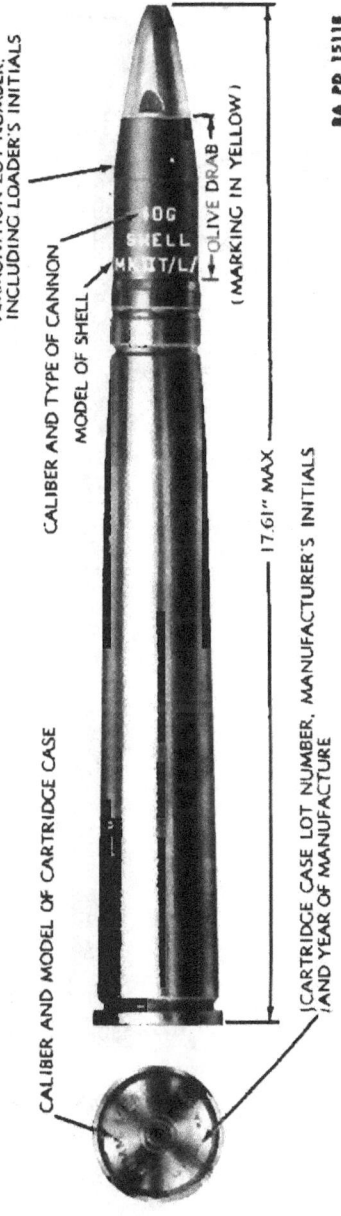

Figure 206 — Cartridge, HE-T (SD, M3), Mk. II, w/ FUZE, P.D., M27 (Navy), 40-mm AA Guns

TM 9-252
111

**40-MM AUTOMATIC GUN M1 (AA) AND 40-MM ANTIAIRCRAFT GUN CARRIAGES M2 AND M2A1**

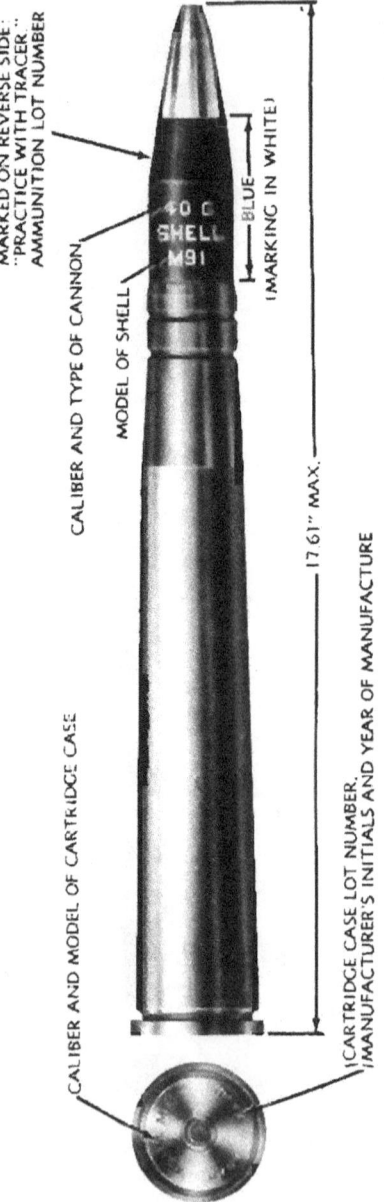

Figure 207 — Cartridge, TP-T, M91, w/FUZE, Dummy, M69, 40-mm AA Guns

262

## AMMUNITION

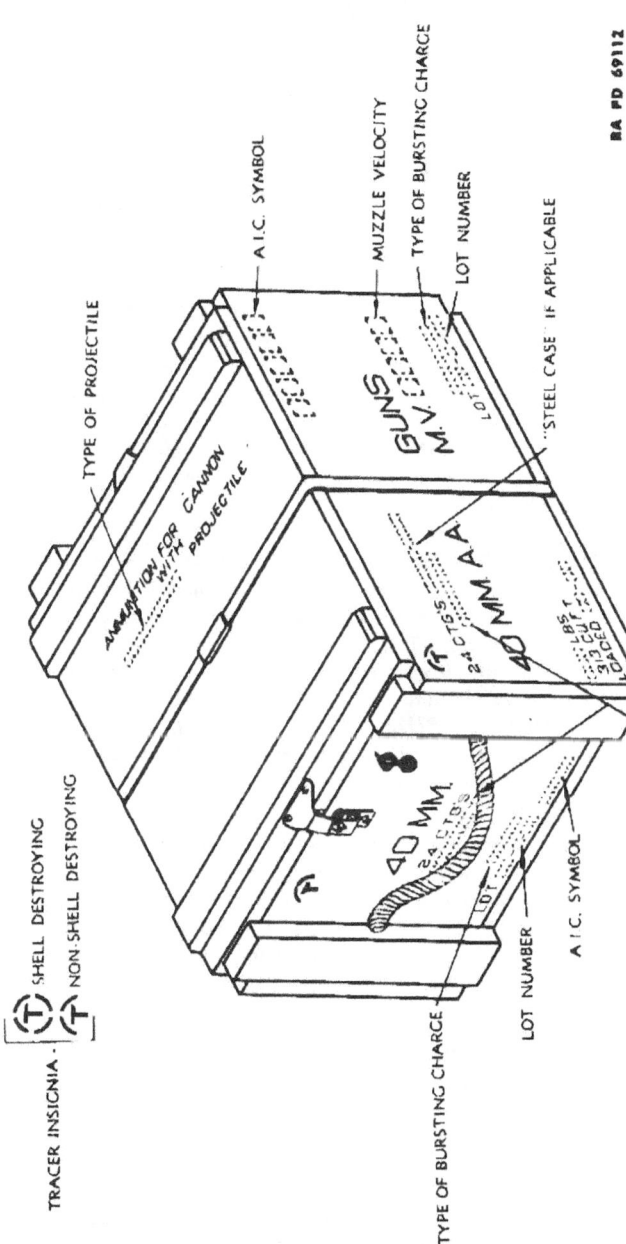

*Figure 208 — Marking on Packing Box for Identification*

TM 9-252
111-113

## 40-MM AUTOMATIC GUN M1 (AA) AND 40-MM ANTIAIRCRAFT GUN CARRIAGES M2 AND M2A1

PRIMER, percussion, M38A1. Ammunition manufactured according to British design will be assembled with CASE, cartridge, M22, an adaptation of the British design. This case differs from the American standard case in that the primer hole is threaded to accommodate the threaded head of the British-type percussion primer, PRIMER, percussion, Q.F., cartridges, No. 12, Mk. II/L.

112. FUZES.

a. General. A fuze is a mechanical device used with the projectile to explode it at the time and under the circumstances desired. Fuzes which function on impact with a very light materiel target such as an airplane wing are known as supersensitive fuzes.

b. Boresafe Fuzes. Dependent upon the manner of arming, certain fuzes are considered to be "boresafe." A boresafe fuze is one in which the explosive train is so interrupted that, prior to firing and while the projectile is still in the bore of the cannon, premature action of the bursting charge is prevented, should any of the more sensitive elements malfunction.

c. FUZE, P.D., Mk. 27 (Navy); FUZE, P.D., M71; FUZE, P.D., M64A1; FUZE, P.D., 251, Mk. I. These fuzes are designed for use with 40-mm antiaircraft ammunition. All are of the impact (point-detonating) type and are issued assembled to the projectile of the fixed complete round. The mechanical elements are so arranged that they function with superquick action on impact with relatively light materiel such as airplane surfaces. The fuzes are considered to be boresafe. However, FUZE, P.D., 251, Mk. I, arms by acceleration whereas FUZE, P.D., M71, and FUZE, P.D., M64A1, arm under rotational forces set up upon firing.

113. PACKING.

a. Data. The complete rounds are packed in two ways, in charger clips holding 4 rounds which in turn are packed in sealed watertight metal containers, and in individual fiber containers in wooden boxes. While weights of the individual rounds vary somewhat, dependent upon the type and model, the following data are considered representative for estimating weight and volume requirements:

| | Weight (lb) | Volume (cu ft) |
|---|---|---|
| In charger clips: | | |
| 4 rounds per charger, 6 chargers (24 rounds) in metal packing box | 151 | 2.22 |
| *Over-all dimensions of box (feet)* 1.72 x 1.48 x 0.87 | | |
| In fiber containers: | | |
| 1 round per container, 24 containers (24 rounds) per wooden box | 159 | 3.12 |
| *Over-all dimensions of box (feet)* 1.83 x 1.51 x 1.13 | | |

## ORGANIZATIONAL SPARE PARTS AND ACCESSORIES

b. Marking for Shipment.

(1) Packings for shipment are marked as follows:

(a) Name and address of destination or port officer (or code marking), preceded by word "To." *

(b) Name and address of ultimate consignee, preceded by word "For." *

(c) List and description of contents. Packing boxes containing ammunition with M3 Tracer will have this model designation stenciled on the box to the right of the tracer insignia.

(d) Muzzle velocity of the rounds contained.

(e) Ammunition Identification Code (A.I.C.) symbol, as published in SNL's.

(f) Gross weight in pounds, displacement in cubic feet.

(g) The number of the package or shipping ticket.*

(h) The letters "U.S." in several conspicuous places.

(i) Order number or contract number.

(j) Ordnance insignium and escutcheon.

(k) Name or designation of consignor preceded by the word "From."*

(l) Lot number.

(m) Month and year packed.

(n) Inspector's stamp.

(2) In addition, the adhesive sealing strips on fiber containers are in the same color as the ammunition item, in accordance with the basic color scheme. It will be noted that for high-explosive projectiles the strips are yellow with black marking.

## 114. FIELD REPORT OF ACCIDENTS.

a. When an accident involving the use of ammunition occurs during training practice, the procedure prescribed in AR 750-10 will be observed by the ordnance officer under whose supervision the ammunition is maintained or issued. Where practicable, reports covering malfunctions of ammunition in combat will be made to the Chief of Ordnance, giving the type of malfunction, type of ammunition, the lot number of the complete rounds or separate-loading components, and condition under which fired.

## Section XII
## ORGANIZATIONAL SPARE PARTS AND ACCESSORIES

### 115. ORGANIZATIONAL SPARE PARTS.

a. A set of spare parts is supplied to the using arms for field replacement of those parts most likely to become broken, worn, or

---

* — May be omitted on individual package in carload shipments of packages of standard weights and dimensions containing standard quantities.

# 40-MM AUTOMATIC GUN M1 (AA) AND 40-MM ANTIAIRCRAFT GUN CARRIAGES M2 AND M2A1

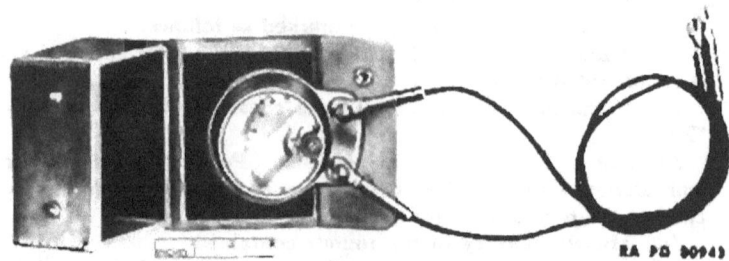

*Figure 209 — Ammeter (for Electric Brakes) A162447*

otherwise unserviceable. The set should be kept complete at all times by requisitioning new parts for those used. Try each part as soon as practicable after received, to see that it fits the materiel properly. For listing of organizational spare parts for the 40-mm gun and carriage, see SNL A-50.

b. Care of spare parts is covered in section VII of this manual.

### 116. ACCESSORIES.

a. Accessories include the tools and equipment required for such disassembling and assembling as the using arms are authorized to perform, and for cleaning and preserving the gun and carriage. Accessories should not be used for purposes other than those prescribed, and when not in use should be properly stored.

b. There are a number of accessories, the names or general characteristics of which indicate their use. Others embodying special features or having special uses, are described in the following paragraphs.

### 117. ACCESSORIES FOR 40-MM GUN AND CARRIAGE.

a. Ammeter (for Electric Brakes). This ammeter (fig. 209) is used for measuring the current taken by the electric brakes of the carriage. It is used by connecting one terminal of the meter to one wire, and the other terminal of the meter to the point to which the wire is normally attached. When the current is supplied to the brakes, this meter registers the current flowing through the brakes.

b. Artillery Gun Book. This artillery gun book (O.O. Form 5825) is used for the purpose of keeping an accurate record of the materiel. It must always remain with the materiel regardless of where it may be sent. The book is divided as follows: record of assignment, battery commander's daily gun record, inspector's record of examination, as well as forms to be filled out in case of premature explosion. This book should be in possession of the organization at all times, and its completeness of records and its whereabouts are the responsibility of the battery commander. It must also contain date of issuance of

# TM 9-252
117

## ORGANIZATIONAL SPARE PARTS AND ACCESSORIES

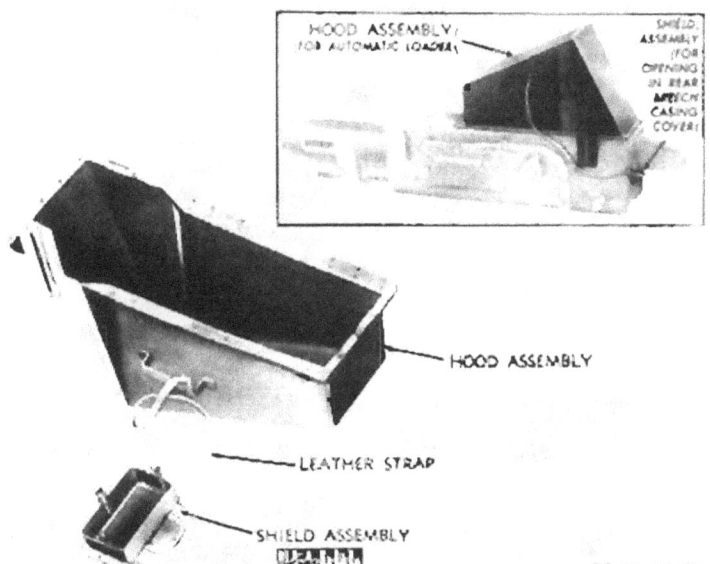

Figure 210 — Automatic Loader Hood and Shield C95025

the materiel, by whom issued, and the place where issued. If a new gun is installed on the carriage, all data recorded in the old book with reference to sights, carriages, etc., must be copied into the new book before the old book is relinquished. Complete instructions on how to make entries in the artillery gun book are contained therein.

NOTE: Record of assignment data must be removed and destroyed prior to entering combat.

c. **Automatic Loader Hood and Shield.** This hood and shield (fig. 210) consists of a large steel hood assembly to which is fastened by a leather strap, a steel shield assembly. It fits over the feed guides and the opening in the rear breech casing cover when the gun is not in use, protecting the automatic loader against dirt and weather.

d. **Barrel Carrier (C95032).** This carrier (fig. 212) consists of two tubular steel handles which are inserted into a cradle. The handles are held in position in the cradle by straight pins that go completely through cradle and handles. The carrier is used to support the rear of the barrel assembly when removing it from and installing it in the breech ring.

e. **Bore Brush M29 (B240944).** This bore brush (fig. 212) consists of a spiral bristle brush with a nut on one end. Its use is to oil the bore of the gun.

f. **Buffer Cylinder Spanner Wrench (A228073).** This wrench

TM 9-252
117

## 40-MM AUTOMATIC GUN M1 (AA) AND 40-MM ANTIAIRCRAFT GUN CARRIAGES M2 AND M2A1

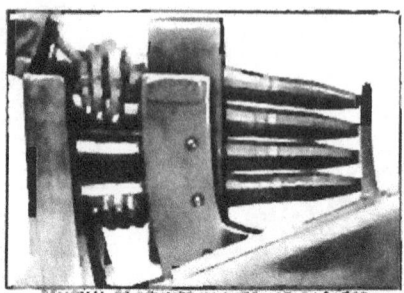

*Figure 211 — Cartridge Remover B200491*

(fig. 212) is constructed of steel having an over-all length of 6.75 inches. Its use is disassembly and reassembly of buffer cylinder on the under side of the loading tray.

g. **Cartridge Remover.** This remover (fig. 211) is used to remove cartridges from the automatic loader. The remover is U-shaped in appearance. When in use, it is pressed over the cartridges and forces the feed and stop pawls of the automatic feed loader device clear, allowing removal of the cartridges. The remover and cartridges are removed together, out of the top of the loader, the pawls being depressed clear.

h. **Cleaning Staff M14 (C97385).** The cleaning staff, 144 inches long (fig. 212), consists of three wooden sections. It is used with the bore brush for cleaning bore of gun. The bore brush is removed and the shell ejector substituted when removing stuck shells from gun chamber.

i. **Closing Spring Cover Wrench.** This wrench (fig. 213) consists of a body, handle, rivet, and spline. These parts when assembled constitute a T-shaped wrench used to remove the cover from the breech ring closing spring assembly.

j. **Electric Brake Control Set M2 (C121778).** This set is supplied with the gun carriage and utilized when the gun carriage is

## ORGANIZATIONAL SPARE PARTS AND ACCESSORIES

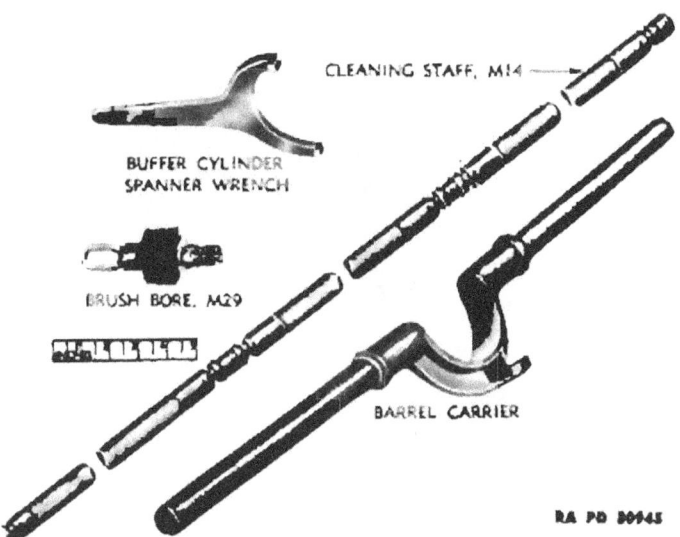

*Figure 212 — Accessories for 40-mm Gun and Carriage*

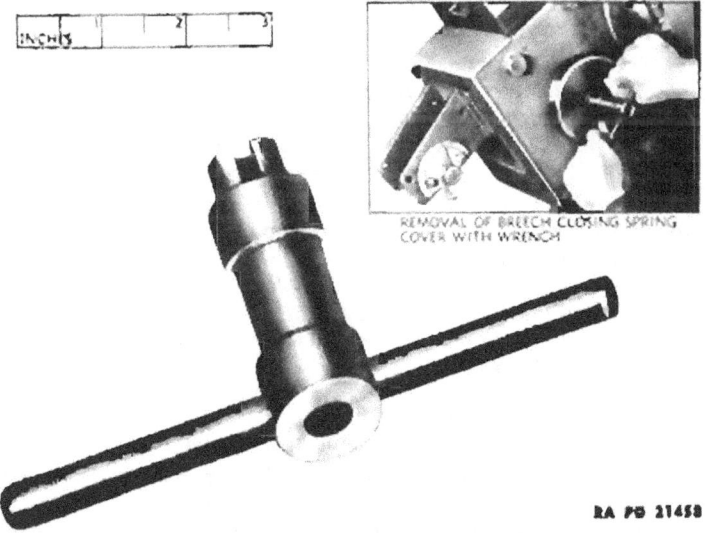

*Figure 213 — Closing Spring Cover Wrench A228074*

# TM 9-252
## 117

### 40-MM AUTOMATIC GUN M1 (AA) AND 40-MM ANTIAIRCRAFT GUN CARRIAGES M2 AND M2A1

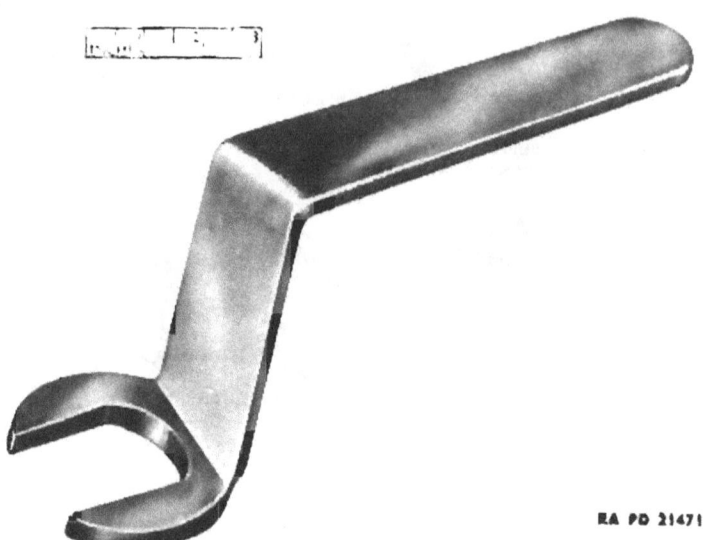

*Figure 214 — Equilibrator Rod Nut Wrench B198554*

towed by a 2½-ton truck (prime mover). This set is operated by the driver of the truck.

k. *Equilibrator Rod Bushing Nut Wrench.* This single-end wrench (fig. 214) which has a handle with two right-angle bends, is used to adjust the compression of the equilibrator springs, also to remove and replace the equilibrator rod jam nut and bushing nut.

l. *Equilibrator Spring Compressor and Guard.* The equilibrator spring compressor (fig. 215) is composed of a threaded shaft, pin clip assembly, guard, and a wrench to which is assembled a bearing, a screw, and two collars. One end of the shaft is threaded and is smaller than the body of the shaft so that it may be screwed into the end of the equilibrator rod assembly with a wrench used on the opposite (shoulder) end of shaft. The shaft is locked in place with the pin clip assembly. The equilibrator spring is compressed by screwing up the wrench on the shaft. The guard, which is inserted through holes in the equilibrator case, is used to hold the compression of the spring after the equilibrator case cover has been opened and the two equilibrator rod assembly nuts have been removed prior to the insertion of the spring compressor. Compression of the spring may be relieved by unscrewing the wrench on the shaft.

m. *Extractor Spindle and Safety Plunger Key Puller.* This puller (fig. 216) is constructed from ⅜-inch bar steel. It has an over-all

## ORGANIZATIONAL SPARE PARTS AND ACCESSORIES

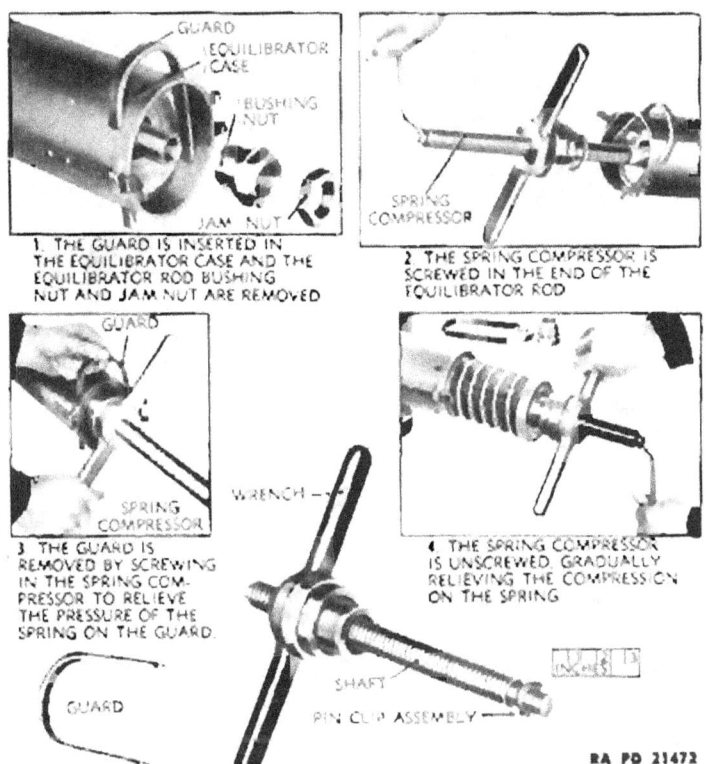

Figure 215 — Equilibrator Spring Compressor C95677

length of 7½-inches. It is used to lift the breech ring safety plunger during the removal of the breechblock and cranks. It is also used for insertion and removal of the extractor spindle from the breech ring.

n. Flash Guard Wrench. This wrench (fig. 217) consists of two tubular handles, two jaws, a pin, and catch assembly. To use, release the catch by pressing downward. This allows the jaws to rotate on the pin. When open, place on the rear hexagonal portion of the flash hider, close jaws, which automatically fasten the catch. The wrench is then in position for removal of the barrel or flash hider.

o. Front and Rear Loader Lifters. These two lifters (fig. 218) are of all metal welded construction. The larger of these two lifters is hooked to the rear portion of the automatic loader, the other to the fore part. When in this position, two men can remove the automatic loader assembly from the gun.

## 40-MM AUTOMATIC GUN M1 (AA) AND 40-MM ANTIAIRCRAFT GUN CARRIAGES M2 AND M2A1

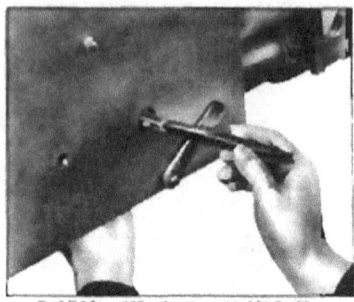

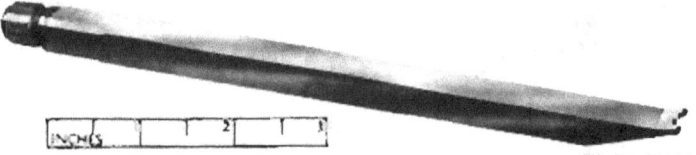

*Figure 216 — Extractor Spindle and Safety Plunger Key Puller A228093*

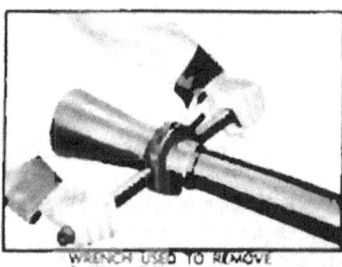

*Figure 217 — Flash Guard Wrench C95038*

# TM 9-252
## ORGANIZATIONAL SPARE PARTS AND ACCESSORIES

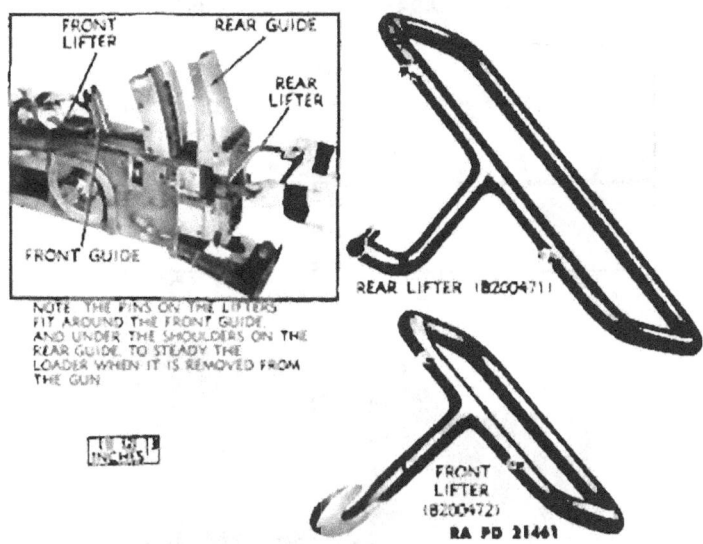

*Figure 218 — Front B200472 and Rear B200471 Loader Lifters*

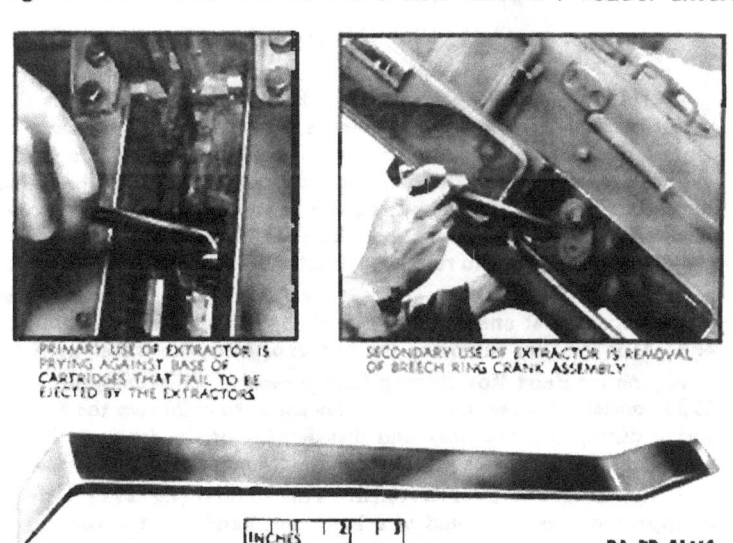

*Figure 219 — Hand Cartridge Extractor A228060*

## 40-MM AUTOMATIC GUN M1 (AA) AND 40-MM ANTIAIRCRAFT GUN CARRIAGES M2 AND M2A1

*Figure 220 — Shell Ejector A298763*

 *p.* **Hand Cartridge Extractor.** This extractor (fig. 219) is of all metal construction having an over-all length of 13.43 inches. The end of the tool bent at approximately a 20-degree angle is used to extract cartridges from the barrel should cartridges fail to be ejected by the extractors. Raise the top cover of the breech casing to accomplish afore-mentioned extraction. The end formed at right angles to the major part is used to remove the outer breech ring crank assembly.

 *q.* **Hub Cap Wrench.** This wrench (fig. 222) has an octagonal shaped opening at one end and a 30-degree offset handle at the other end. It is used for removing the hub caps from the wheel hubs.

 *r.* **Loader Feed Rod Spring Compressor.** This compressor (fig. 222) consists of a jaw and screw. Its use is to compress the feed rod spring during the assembly and disassembly of the feed rod holders and feed rods to replace a broken or weak loader feed rod spring.

 *s.* **Locking Pin Seat Wrench.** This wrench (fig. 222) T-shaped in appearance, is composed of a body and handle. It is used to unscrew spring seat allowing removal of breech ring safety plunger.

 *t.* **Shell Ejector.** This ejector (fig. 220) is used to remove live shells stuck in the chamber of the gun tube with the rotating band wedged in the origin of rifling. The ejector is bored at one end and shaped to suit the shoulder of the shell and clear the fuze; the other

TM 9-252
117

## ORGANIZATIONAL SPARE PARTS AND ACCESSORIES

SHELL PUSHER WITH FLAT END INSERTED BETWEEN FEED ROLLERS FORCING ROUNDS UPWARD INTO CARTRIDGE REMOVER

SHELL PUSHER IN POSITION FOR REMOVING ROUNDS BY FORCING THEM THROUGH FEED ROLLER ONTO LOADING TRAY

Figure 221 — Shell Pusher C95031

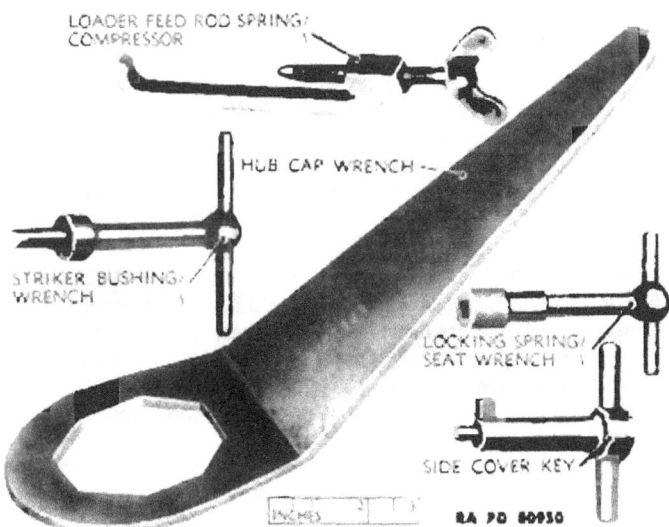

Figure 222 — Accessories for 40-mm Gun and Carriage

275

TM 9-252
117

## 40-MM AUTOMATIC GUN M1 (AA) AND 40-MM ANTIAIRCRAFT GUN CARRIAGES M2 AND M2A1

*Figure 223 — Striker Protrusion Gage A276650*

end has a screw-threaded projection to fit the socket of the cleaning staff. The ejector is screwed on the cleaning staff and inserted into the muzzle of the gun tube. The shell is then pushed out of the gun tube chamber by gradually forcing the cleaning staff-shell ejector assembly to the rear. The hand operating lever should be held fully to the rear in order to depress the breechblock fully. The person removing the stuck shell should stand as much as possible to the side of the muzzle as a safety precaution.

u. **Shell Pusher.** This pusher assembly (fig. 221) consists of a handle, brace, and pusher. The brace is securely welded to the handle approximately 3.5 inches from the pusher, which is fastened by a pin to the working end of the handle. This pusher assembly is used to unload cartridges remaining in the automatic loader at the conclusion of firing. There are two methods by which this may be accomplished:

(1) Force the cartridges between the feed rollers on to the loading tray (with the rammer cocked and the hand operating lever in the rear position) by means of the pusher, using the feed pawls as a fulcrum. The hand operating lever must be rotated to the rear each time a cartridge is forced between the feed rollers; this is necessary to release the feed catches.

(2) Insert the prepared flat end of the pusher into the rear of the

276

automatic loader between the feed rollers, and force the cartridges upward into the cartridge remover (fig. 211). Care must be taken not to use undue pressure when forcing the rounds upward.

v. Side Cover Key. This key (fig. 222) is of all metal construction having a bar handle. Its use is to lock and unlock the side cover. To use, insert into opening provided in side cover, exerting just enough pressure against spring tension to allow turning, thereby releasing catch.

w. Striker Bushing Wrench. This wrench (fig. 222) consists of a body and handle and is used to remove the loading tray bolt spring seat, breechblock firing pin spring cover, and loading tray attaching bolt.

x. Striker Protrusion Gage. This gage (fig. 223) is used to determine if the length of protrusion of the striker or firing pin is within the specified limits (0.114 in. max., 0.099 in. min.). When placing the gage over the firing pin against the face of the breechblock, it can be determined in two steps whether the length of protrusion is less than the minimum allowable 0.099 inch or more than the maximum allowable 0.114 inch. Firing pins that do not fall within these limits should be replaced.

## Section XIII
## STORAGE AND SHIPMENT

### 118. PREPARATION FOR DOMESTIC SHIPMENT.

a. General. The 40-mm Automatic Gun M1 and Carriages M2 and M2A1 (AA) can be shipped and stored either with or without the gun mounted. All precautions should be taken to prevent corrosion during shipment and storage, to keep the recoil mechanism exercised, and to prevent the deterioration of rubber during storage. The materiel should be prepared for both shipment and storage as directed in subparagraph b, below.

b. Preparation of 40-mm Automatic Gun M1 and Carriages M2 and M2A1 (AA).

(1) LUBRICATION. The materiel should be completely lubricated before shipment or storage in acordance with lubrication instructions as directed in section VI.

(2) CLEANING.

(a) The materiel shall be thoroughly cleaned and made free of all foreign matter, using SOLVENT, dry-cleaning, or a soap solution.

(b) Special attention should be given to breech and firing mechanism, and where possible, a partial disassembly of these components should be made to insure thorough cleaning.

## 40-MM AUTOMATIC GUN M1 (AA) AND 40-MM ANTIAIRCRAFT GUN CARRIAGES M2 AND M2A1

(c) Apply SOLVENT, dry-cleaning, by scrubbing with a clean brush or wiping with clean saturated cloths.

(d) Apply soap solution by vigorously brushing or scrubbing the surfaces thoroughly until all traces of contamination have been removed. Rinse the cleaned surfaces with clean, hot water and dry thoroughly.

(e) Avoid contact of bare hands with the cleaned surfaces.

(3) PAINTING. Painted surfaces that have become checked, pitted, or rusted must have the rust spots removed and the surfaces repainted.

(a) *Removing Rust Spots.* The following may be used in removing rust spots:

1. CLOTH, abrasive, aluminum-oxide, for cleaning finished and unfinished external surfaces where wear of the parts cleaned will not affect the functioning of the mechanism.

2. CLOTH, crocus, for removing rust or stain and polishing parts of the breechblock and firing mechanism and other finished surfaces of metal.

(b) *Application of PRIMER, Synthetic, Rust-inhibiting (AXS-750).* Apply a liberal coating of PRIMER, synthetic, rust-inhibiting over entire area of the cleaned surfaces to be repainted as follows:

1. *Brushing or spraying.* PRIMER, synthetic, rust-inhibiting, should be used on bare metal as a base coat for synthetic enamel. It may be applied either by brushing or spraying. The primer will brush satisfactorily as received or after the addition of not more than 5 percent by volume of THINNER, paint, volatile mineral spirits, TT-T-291. For spraying, the primer may be thinned with not more than 15 percent by volume of THINNER, paint, volatile mineral spirits, TT-T-291. Allow to dry throughly.

(c) *Sandpapering Surfaces.* Sandpaper the primed surfaces with PAPER, flint, class B, No. 00, and wipe all particles of dust from surfaces.

(d) *Application of Enamel.* Apply coat of ENAMEL, synthetic, olive-drab, lusterless, and allow to dry thoroughly before the materiel is used.

(4) APPLICATION OF PRESERVATIVE. NOTE: Application of preservatives should be accomplished immediately after cleaning and drying. Rust-preventive compounds, light and heavy, used herein shall be brought to the proper consistency by heating before application.

(a) Apply a coating of COMPOUND, rust-preventive, light, to interior portions of breech mechanism. Apply a coating of COMPOUND, rust-preventive, heavy, to all exterior portions of the breech.

(b) Thoroughly swab the bore of the gun with COMPOUND, rust-preventive, light.

(c) *External Unpainted Surfaces.*

## STORAGE AND SHIPMENT

*1.* Use COMPOUND, rust-preventive, thin film, on external unpainted surfaces of the carriage that are not highly finished, critical, or operating, and from which preservative *need not be completely removed* before materiel is placed in operation. This compound will be applied by brushing or spraying without heating or solvent dilution.

*2.* Use COMPOUND, rust-preventive, heavy, on external unpainted, critical operating surfaces of the carriage from which preservative *must be completely removed* before materiel is placed in operation.

(5) COVERS. Install breech and muzzle covers supplied with the materiel, and fasten securely. Using TAPE, adhesive, nonhygroscopic, wrap that portion of tube adjacent to the end of muzzle cover and over the end of cover, sealing the muzzle end of the gun. If covers are not available, the muzzle and breech ends of the gun shall be sealed using two layers of PAPER, greaseproof, wrapping, and one of PAPER, Kraft, wrapping, waterproofed. Apply TAPE, adhesive, nonhygroscopic, over paper completely sealing the openings.

(6) MISCELLANEOUS. Free end of electric brake jumper cable and coupling shall be taped and tied securely to the drawbar.

(7) GENERAL INSPECTION. Make a systematic inspection just before shipment or storage, and list all broken or missing items that are not repaired or replaced, and attach this list to the materiel.

## 119. LOADING AND BLOCKING MATERIEL ON RAILROAD CAR.

a. General. All loading and blocking instructions as specified herein are minimum, and are in accordance with the Association of American Railroads, "Rules Governing the Loading of Commodities on Open Top Cars, Special Supplement, Revised 1 March 1943," containing the "Rules Governing the Loading of Mechanized and Motorized Equipment and Major Calibre Guns."

b. Instructions.

(1) INSPECTION. Railroad cars must be inspected to see that they are suitable to carry loads to destinations. Floors must be sound and all loose nails or other projections not an integral part of the car should be removed.

(2) RAMPS. Permanent ramps should be used for loading the materiel when available, but when such ramps are not available, improvised ramps may be constructed of rail ties and other available lumber.

(3) HANDLING.

(a) Cars loaded in accordance with specifications given herein must not be handled in hump switching.

(b) Cars must not be cut off while in motion, and must be coupled carefully and all unnecessary shocks avoided.

(c) Cars must be placed in yards or sidings so that they will be subjected to as little handling as possible. Separate track or tracks,

TM 9-252
119
40-MM AUTOMATIC GUN M1 (AA) AND 40-MM ANTIAIRCRAFT
GUN CARRIAGES M2 AND M2A1

when available, must be designated at terminals, classification or receiving yards, for such cars, and cars must be coupled at all times during such holding and hand brakes set.

(4) PLACARDING. Materiel not moving in combat service must be placarded, "DO NOT HUMP."

(5) CLEARING LIMITS. The height and width of load must be within the clearance limits of the railroads over which it is to be moved. Army and railroad officials must check all clearances prior to each move.

(6) MAXIMUM LOAD WEIGHTS. In determining the maximum weight of load, the following shall govern, except where load weight limit has been reduced by the car owner:

| Marked Capacity of Car (Pounds) | Total Weight of Car Load (Pounds) | Load Weight (lb. wt. of car to be deducted) |
|---|---|---|
| 40,000 | 66,000 | 66,000 |
| 60,000 | 103,000 | 103,000 |
| 80,000 | 136,000 | 136,000 |
| 100,000 | 169,000 | 169,000 |
| 140,000 | 210,000 | 210,000 |
| 200,000 | 251,000 | 251,000 |

EXAMPLE:
Capacity of car ................................................................ 100,000 lb
Total weight of car and load ........................................... 169,000 lb
*Light weight of car (to be subtracted) ............................ 37,000 lb
Permissible weight of load .............................................. 132,000 lb

NOTE: When loading railroad cars, materiel shall be so loaded as to require a minimum number of cars. To accomplish this, various types of materiel may be loaded on the same car provided all have the same destination.

(7) BRAKE WHEEL CLEARANCE (A, fig. 224). Each railroad car must be loaded with a resulting brake wheel clearance of at least 6 inches in front, at each side and at the top. Brake wheel clearance shall be increased as much as is consistent with proper location of load.

(8) BRAKES. After loading and bracing the materiel, set the hand brakes.

(9) TIRES. Tire pressure shall be increased to 55 pounds per square inch.

(10) TYPE OF CARS. Flat, box, or drop end gondola cars may be used.

(11) DRAWBAR REMOVAL. Drawbar should be removed and secured to floor of railroad car by strapping with No. 8 gage, black annealed wire or flat steel strapping. Attach wire to floor by wrapping around blocks and nailing the blocks to the floor. Secure flat steel

---
*This marking is stenciled on each side of car indicated as "LT.WT."
Load must be so placed on the car that there will not be more weight on one side of the car than on the other. One truck of the carrying car must not carry more than one-half of the load weight.

## STORAGE AND SHIPMENT

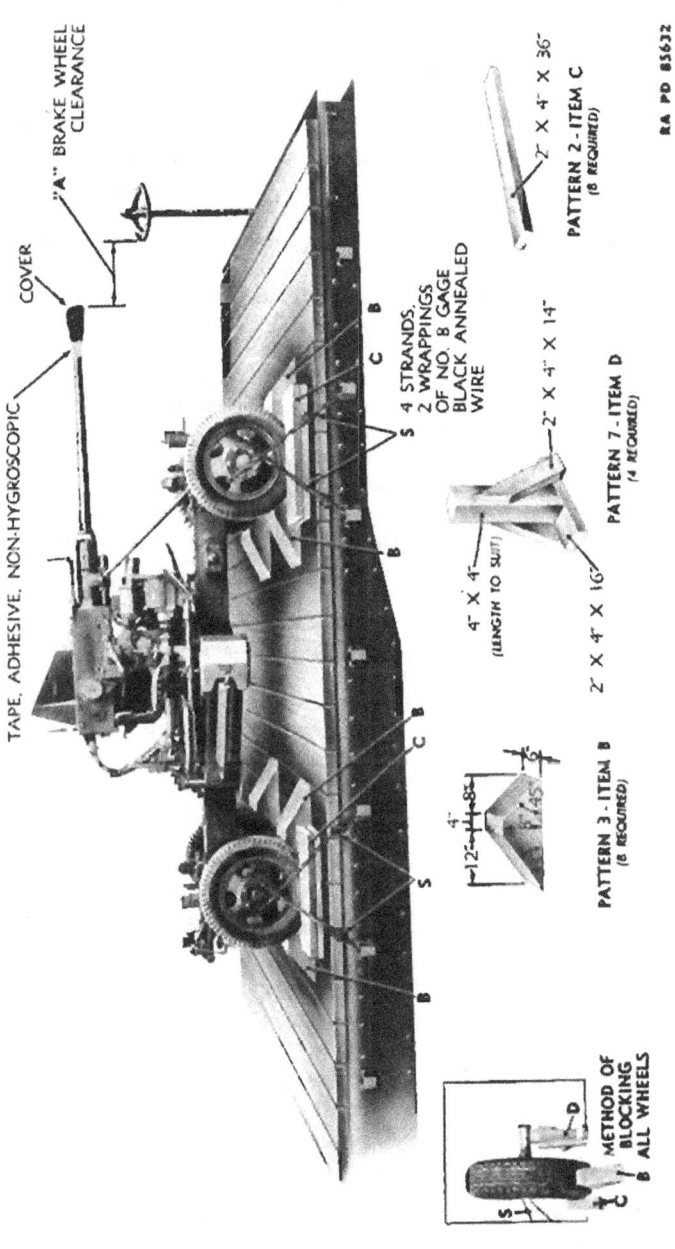

Figure 224 — Method of Blocking 40-mm Automatic Gun M1 and Carriages M2 and M2A1 (AA) on Railroad Car

## 40-MM AUTOMATIC GUN M1 (AA) AND 40-MM ANTIAIRCRAFT GUN CARRIAGES M2 AND M2A1

strapping that is not punched for nailing to floor by means of anchor plates attached to the car floor.

(12) GUN POSITION. Place gun in its traveling position with the gun stay in place.

(13) OUTRIGGER COVERS. Outriggers should be positioned for traveling with the covers in place over the hinge joints for protection from dirt and weather.

c. Blocking. All item reference letters given below refer to the details and locations in figure 224.

(1) PATTERN 3, ITEM B (EIGHT PIECES OF PATTERN 3 REQUIRED). Place one pattern 3 at the front and one at the rear of each wheel. The 45-degree portion of the block shall be placed next to the tire. Nail the heel of the pattern to the car floor with three 40-penny nails and toenail that portion under tire to car floor with two 40-penny nails before patterns 2 are applied.

(2) PATTERN 2, ITEM C (EIGHT PIECES OF PATTERN 2 REQUIRED). Place two patterns 2 on each side of the carriage against the outside of each tire. Nail the bottom pattern to the car floor with three 40-penny nails and the top pattern to the pattern below with three 40-penny nails.

(3) PATTERN 7, ITEM D (FOUR PIECES OF PATTERN 7 REQUIRED). Place one pattern 7 under the axle near the inside face of each wheel. These patterns should be cut so as to fit snugly between the car floor and the axle to relieve partially the weight from the tires. Nail each pattern to the car floor with six 40-penny nails.

(4) STRAPPING, ITEM S. Secure each wheel by passing wire consisting of four strands, two wrappings of No. 8 gage, black annealed wire through two openings in the wheels and securing the wire to the stake pockets on each side of the car. Twist tie with rod or bolt enough to remove slack. The openings in the wheels through which the wire passes should be approximately the same distance from the car floor. NOTE: When drop end gondola car is used, apply strapping in similar fashion, and attach to the floor by use of blocking or anchor plates.

### 120. LIMITED STORAGE INSTRUCTIONS.

a. When the materiel is stored uncrated, preparation will be in accordance with paragraph 118 b.

b. Periodical Inspections. Periodical inspections shall be made while the materiel is stored, to note among other things, general condition, missing parts, and the need for repairs. If found to be corroding at any part, the entire procedure as given herein under paragraph 118 b will be repeated.

c. Inspect the tires, repair any leaks that have developed, and inflate, if necessary.

## Section XIV

## OPERATION UNDER UNUSUAL CONDITIONS

### 121. GENERAL.

a. Because of the different climates in which this materiel may be expected to operate, special instructions are given in this section for three regions, namely: Arctic, temperate, and tropical.

b. In certain cases, the prescribed instructions may not apply; for example, a tropical climate may be experienced in a temperate region. In cases of this nature, the instructions as to the classification of the climate in which the materiel is operating is left to the judgement of the ordnance officer. He is cautioned, however, that only extended, and not temporary, periods of climatic conditions govern the classification.

c. Manufacturing arsenals and plants should lubricate the materiel on assembly as prescribed in the lubrication guides (figs. 104 and 105). If the materiel is to be used in a climate other than temperate, the precautions in paragraphs 119 and 120 should be taken.

d. Materiel, previously lubricated for a colder climate than the one in which the materiel is to be used, should be relubricated with lubricants prescribed for use in that climate.

e. Materiel, previously lubricated for a warmer climate than the one in which the materiel is to be used, should be completely cleaned of all lubricants and relubricated with the lubricants prescribed for use in that climate.

### 122. TROPICAL CLIMATES.

a. The bore of the gun should be cleaned and oiled more frequently than usual when operating in hot climates. Temperature changes will cause condensation of moisture in the air on metal, and cause rusting. If condensation occurs on other metal parts of the gun and carriage, wipe them dry and coat with oil as required to prevent rusting.

b. Materiel should be inspected frequently when being operated in hot, moist areas. Cloth covers and other items which may deteriorate from mildew, or be attacked by insects or vermin should be aired and dried often.

c. Lubrication. Lubricate the materiel and check the levels of lubricants in gear cases more frequently when operating in hot climates. Heat will cause lubricants to thin out and deteriorate more quickly; gear cases and other lubricant-retaining devices may leak when the lubricant is hot. Use the lubricants prescribed for temperatures over 32 F as instructed in section VI.

d. Tires. Tires should be checked at frequent intervals to ascertain that the pressure is not above that which is prescribed, 45 pounds per square inch. Tires must be kept out of the direct rays of the sun

# 40-MM AUTOMATIC GUN M1 (AA) AND 40-MM ANTIAIRCRAFT GUN CARRIAGES M2 AND M2A1

as much as possible. They should be covered to protect them from prolonged exposure to the sun.

*c.* **Ammunition.** Ammunition should be kept cool, and always out of the direct rays of the sun. Moisture-resistant seals should not be broken until ammunition is to be used.

NOTE: Extreme heat is often accompanied by other adverse conditions. Refer to paragraphs 121 and 122.

## 123. ARCTIC (SUB-ZERO) CLIMATES.

*a.* When the weapon is operated at sub-zero temperatures, special precautions must be taken in its care and maintenance to avoid poor performance and/or total functional failure, and in some instances, damage to both materiel and personnel. The materiel can be operated efficiently if the instructions given here are followed, even when average temperatures below zero degree F prevail. These instructions also should be followed when sluggish or stiff operation of the materiel indicates the advisability of adopting them at higher temperatures.

*b.* **Introduction to Low Temperature Lubrication.**

(1) Lubricating oils stiffen progressively as temperature drops until a point is reached at which they solidify. As stiffness increases, the power required to move bearing surfaces and gears in contact with the oil multiplies rapidly until movement becomes impossible. When solidification occurs, the moving parts cut a channel through the solid oil leaving the rubbing surfaces dry and unlubricated. Before friction can develop enough heat to liquefy the oil and reestablish an oil film, bearing and gear tooth surfaces may score and fail completely.

(2) Similar action takes place where rubbing surfaces are fed by an oil pump. The stiffened oil flows too slowly, or not at all, to the pump inlet and the oil already in the feed lines cannot be forced to the bearings. Oils, prescribed for use in artillery materiel at high temperatures, are designed to maintain adequate body at those temperatures. In most instances, they become too stiff at low temperatures for satisfactory operation. They must therefore be completely replaced in cold weather with lighter oils which will remain fluid at the lowest expected operating temperatures.

(3) Grease is a combination of soap and oil. The soap acts as a sponge to hold the oil in place, thus preventing leakage which might occur if oil alone were used. As temperature decreases, both oil and soap stiffen and retard the movement of parts. In cold weather, therefore, greases which cause minimum drag must be used.

(4) The presence of only a small quantity of warm weather grease may immobilize bearings and prevent the manipulation of the gun at sub-zero temperatures. When sub-zero temperatures are expected, all grease used in temperatures above zero degrees F must be removed

**TM 9-252**
**123**

### OPERATION UNDER UNUSUAL CONDITIONS

from bearings and gear cases and replaced with the specified low temperature products. Once grease has solidified, it cannot be removed from bearings or gear cases without the application of enough heat to melt it, or the disassembly of the unit and the washing of the parts with SOLVENT, dry-cleaning.

c. **Mechanical Condition.** Since metals contract and expand with temperature change and the amount of expansion and contraction varies with different metals, the clearance between bearing surfaces consisting of dissimilar metals is considerably less at sub-zero than at higher temperatures. In preparing artillery for sub-zero operation, therefore, care should be taken that parts are alined properly and normal clearances exist. This applies not only to bearings but also to mechanisms employing packings around rotating or reciprocating shafts or rods. Lack of attention to this may result in binding which will make mechanisms stiff or inoperative, regardless of the lubricant used. Scored or rough bearing surfaces also interfere with easy action and should be smoothed in preparing materiel for low temperature operation.

d. **Cleanliness.** Cleanliness is imperative. Rust, dirt, and gummy oil and grease in bearing clearances reduced by low temperature will interfere with the proper feed of lubricant. This will cause stiff action, if not complete stoppage. In *winterizing* materiel, therefore, assemblies and mechanisms must be disassembled sufficiently to permit thorough removal of old oil, grease, and foreign matter. Cleaning is most efficiently done by washing with SOLVENT, dry-cleaning, employing brushes and scrapers where necessary. Field experience has proved that neglect in cleaning small linkages, bearings, and other similar parts may cause malfunctioning and stoppages in sub-zero weather.

e. **Timeliness.** Placing of materiel in proper mechanical condition requires time for necessary disassembly, repair, and cleaning, and must be carefully done. The approach of low temperature must be anticipated far enough in advance to permit completion of the conditioning before the onset of sub-zero temperatures.

f. **Prescribed Lubricants.** The following materials will be used for the lubrication and preservation of the gun and carriage at temperatures below zero degree F, and above zero degree F where stiff action indicates their necessity. Refer to SNL K-1 for specification designations of products available through the ordnance provision system.

(1) OIL, LUBRICATING, PRESERVATIVE, LIGHT. All bearings, gear cases, and mechanisms for which the prescribed oil at normal temperatures is engine oil. Miscellaneous hand oiling.

(2) OIL, LUBRICATING, FOR AIRCRAFT INSTRUMENTS AND MACHINE GUNS. Oil-lubricated parts in directors, sighting instruments, and other fire control equipment.

TM 9-252
123

### 40-MM AUTOMATIC GUN M1 (AA) AND 40-MM ANTIAIRCRAFT GUN CARRIAGES M2 AND M2A1

(3) OIL, HYDRAULIC. All hydraulic systems for speed gears and remote control oil gears.

(4) GREASE, O.D., No. 00. All bearings, gears, and mechanisms for which GREASE, O.D., No. 0, is prescribed at temperatures above 32 F.

(5) GREASE, LUBRICATING, SPECIAL. Grease-lubricated mechanisms of fire control materiel.

(6) GREASE, GENERAL PURPOSE, No. 2. Wheel bearings at all temperatures.

(7) CLEANER, RIFLE BORE. May be used for cleaning gun bores after firing if a cleaning solution is not available.

(8) SOLUTIONS, SPONGING. Cleaning gun bores after firing.

(9) SOLVENT, DRY-CLEANING. Cleaning grease and oil from all mechanisms and parts. May also be used for diluting OIL, engine, SAE 10, in engine crankcases.

g. Low Temperature Lubricating Instructions. To insure adequate lubrication and satisfactory performance of artillery materiel in cold weather, the following instructions must be followed when sub-zero temperatures are expected:

(1) BALL AND ROLLER BEARINGS, GREASE LUBRICATED. It is impossible to replace warm weather grease in ball and roller bearings by forcing in the grease prescribed for low temperatures. Attempts to do this result in overloading the bearings with unsuitable lubricant which will congeal at low temperatures and immobilize the moving parts. These bearings must be removed by disassembly from the materiel, washed thoroughly in SOLVENT, dry-cleaning, to remove all traces of heavy grease, dried, and then repacked with the prescribed lubricant in accordance with instructions in War Department Lubrication Guide, No. 61.

(2) BALL AND ROLLER BEARINGS, OIL-LUBRICATED. Oil-lubricated ball and roller bearings preferably should be removed and cleaned. If this is impractical, a thorough flushing with SOLVENT, dry-cleaning, followed by application of the prescribed oil generally will give satisfactory results. Oil sumps and reservoirs must be drained, flushed, and filled with the proper oil. The wicks of wick-fed bearings should be removed and saturated with OIL, lubricating, preservative, light, before reassembling.

(3) PLAIN JOURNAL BEARINGS AND BUSHINGS. It is preferable to disassemble these bearings, not only to remove thoroughly all heavy oil and grease, but also to smooth all roughness and to test for adequate clearance between shaft and bearing. Where disassembly is impracticable, heavy lubricant usually can be forced from the bearings by thorough flushing with OIL, lubricating, preservative, light. Reservoirs and wick feeds must be cleaned completely and

**TM 9-252**
**123**

## OPERATION UNDER UNUSUAL CONDITIONS

refilled to the prescribed level with the proper oil.

(4) GEARS, OIL-LUBRICATED. Where gears are inclosed in oiltight gear cases filled with oil, all oil will be drained, the case flushed with SOLVENT, dry-cleaning, and the case refilled to the proper level with OIL, lubricating, preservative, light. Where no drain or level plug is provided, the gear cases will be disassembled by ordnance maintenance personnel.

CAUTION: Do not fill above the prescribed level as the surplus oil will cause unnecessary drag on the movement of the gears.

(5) GEARS, GREASE-LUBRICATED. Since it is practically impossible to wash heavy grease out of a gear case by flushing, grease-filled cases will be disassembled, cleaned, and refilled by ordnance maintenance personnel.

h. Elevating Arc. It is often found that snow will collect on the elevating arc. This snow will cake under the pressure of the gears, forming ice and interfering with the elevation of the piece. This snow must be removed by vigorous brushing with a stiff bristle or wire brush before elevation of the piece is attempted.

i. Gun Bore.

(1) Cleaning of a gun bore after firing cannot be accomplished in the normal manner at temperatures below 32 F because the water will freeze in the tube. If the cleaning can be done with the tube hot, and hot water is available, normal SODA ASH or soap solutions can be used. Otherwise, it will be necessary to add alcohol, glycerine, or COMPOUND, antifreeze, to the normal SODA ASH or soap solution to prevent freezing.

(2) To ten parts by volume of cleaning solution, add the number of parts of one of the antifreezes shown below:

| Temperature F | Glycerine | or Alcohol | or Ethylene Glycol |
|---|---|---|---|
| 20 | 2½ | 2 | 2 |
| 10 | 5 | 4 | 3½ |
| 0 | 6½ | 6½ | 5 |
| −15 | 10 | 9 | 7½ |
| −30 | 13 | 16 | 10 |
| −40 | 16 | 27 | 12 |

(3) When available, CLEANER, rifle bore, may be used. However, this solution will freeze at temperatures below 32 F. If frozen, it must be thawed and shaken well before using. Closed containers should not be filled to more than 75 percent of capacity in freezing weather. Completely filled containers will burst when contents freeze. In an emergency, SOLVENT, dry-cleaning, or OIL, lubri-

TM 9-252
123

**40-MM AUTOMATIC GUN M1 (AA) AND 40-MM ANTIAIRCRAFT GUN CARRIAGES M2 AND M2A1**

cating, preservative, light, may be used, but neither is as effective as the cleaning solution.

(4) In applying OIL, lubricating, preservative, light, to the bore after cleaning, care must be taken to work the oil in well so that it will reach all surfaces of the lands and grooves. When the gun is brought into a heated shop, condensation will occur on all metal surfaces. After the gun reaches shop temperature, the tube and all other bright metal parts must be wiped dry and recoated with oil to prevent rusting.

j. *Firing Pin and Attendant Parts, Breechblock and Breech Casing Firing Mechanism Parts.* Not only must extreme cleanliness be maintained, but oil must be applied sparingly to obtain proper functioning in cold weather. The best method of application, after cleaning thoroughly with SOLVENT, dry-cleaning, is to wipe the rubbing surfaces of the parts with a clean cloth which has been wet with oil and *wrung out*. Use OIL, lubricating, preservative, light, for normal winter temperatures. In extreme cold, use OIL, lubricating, for aircraft instruments and machine guns.

k. *Recoil Cylinder.* The recoil cylinder is filled with a mixture of 60 parts by volume of OIL, hydraulic, and 40 parts by volume of OIL, recoil, light. This mixture will be used in all temperatures.

l. *Oil Gears.* Although hydraulic oil is considered satisfactory for sub-zero temperatures, it may be expected to stiffen to some extent at low temperatures. Oil gears should be exercised for periods of at least 30 minutes, often enough to assure that the materiel functions properly and is ready for action at all times. This exercising will warm the oil and gears.

m. *Sighting and Fire Control Equipment.* Sighting and fire control equipment is normally lubricated for operation over a wide range of temperatures. This equipment should be exercised frequently during periods of low temperature to insure proper functioning. If any equipment does not function properly, the ordnance maintenance personnel should be notified.

n. *Emplacement.* Gun carriages and supports for fire control equipment freeze quite solidly into the ground in cold weather. As a result, considerable labor is required to dig the equipment if the battery position must be changed. To prevent this, coat all surfaces that contact the ground with waste grease or very heavy oil before emplacing. Sources of waste lubricant are the gear casings of guns and automotive equipment, the drainings and scrapings of these being used. They are obtained when the materiel is relubricated for cold weather.

o. *Electric Brakes.* Before traveling, the safety switch should always be checked for proper operation because low temperatures

## OPERATION UNDER UNUSUAL CONDITIONS

seriously reduce the efficiency of the "hot-shot" batteries. Partly discharged batteries must be replaced with new and fully charged units. The plugs must be kept clear of ice and snow or it will be impossible to insert them in their sockets.

p. *Tires.* Tires should be checked at frequent intervals to insure that the pressure is not below that prescribed, 45 pounds per square inch.

q. *Dry Cell Batteries.* Dry cell batteries are very unsatisfactory after exposure for a few hours to low temperatures.

r. *Inspection.* Inspections should be made daily because of the limited amount of lubricant used in cold weather operations.

s. *Operation.* Keep the loading platform free of ice or snow to prevent injury to personnel caused by insecure footing. The bore should be inspected frequently for frosting or congealed lubricants. The flash hider cover and the automatic loader hood and shield should be in place as much of the time as the rate of fire permits to exclude snow and moisture from the interior of the weapon.

## 124. EXCESSIVELY MOIST OR SALTY ATMOSPHERE.

a. When the materiel is active, clean and relubricate the bore and exposed metal surfaces more frequently than is prescribed for normal service. Moist and salty atmospheres have a tendency to emulsify oils and greases and destroy their rust-preventive qualities. Inspect parts frequently for corrosion. Keep the flash hider cover and the automatic loader hood and shield in place as much of the time as firing conditions permit.

b. Cloth covers and other items that may deteriorate from dampness should be inspected frequently and dried as often as possible.

c. When the materiel is inactive, the unpainted parts should be covered with a film of COMPOUND, rust-preventive, heavy, and should be inspected daily for traces of the formation of rust. All covers should be in place.

## 125. EXCESSIVELY SANDY OR DUSTY CONDITIONS.

a. When the weapon is active in dusty or sandy areas, remove lubricants from the elevating rack and from exposed sliding parts and bearing surfaces. Surface lubricant will pick up sand and dust, forming an abrasive which will cause rapid wear. Lubricate the parts after action.

b. Inspect and lubricate the materiel more frequently when operating in sandy or dusty areas. Exercise particular care to keep sand and dust out of the mechanisms and oil receptacles when carrying out inspecting and lubricating operations and when making ad-

TM 9-252
125

### 40-MM AUTOMATIC GUN M1 (AA) AND 40-MM ANTIAIRCRAFT GUN CARRIAGES M2 AND M2A1

justments and repairs. Cover the cooling slots in the forward portion of the breech casings of those guns having these slots. Seal all openings in the rear portion of the breech casing. Keep all covers in place as much of the time as firing conditions permit. Shield parts from flying sand and dust with paulins or the carriage cover during disassembly and assembly operations.

c. Should the prescribed lubricant for the gun bore prove inadequate in preventing rusting of the bore surfaces, OIL, lubricating, preservative, medium, may be used for oiling of the bore during periods of inactivity.

## Section XV

## REFERENCES

### 126. PUBLICATIONS INDEXES.

The following publications indexes should be consulted frequently for latest changes or revisions of references given in this section and for new publications relating to materiel covered in this manual:

a. Introduction to Ordnance Catalog (explaining SNL system) .................... ASF Cat. ORD 1 IOC

b. Ordnance Publications for Supply Index (index to SNL's) .................... ASF Cat. ORD 2 OPSI

c. Index to Ordnance Publications (listing FM's, TM's, TC's, and TB's of interest to ordnance personnel, OPSR, MWO's, BSD, S of R's, OSSC's, and OFSB's; and includes Alphabetical List of Major Items with Publications Pertaining Thereto) .................... OFSB 1-1

d. List of Publications for Training (listing MR's, MTP's, T/BA's, T/A's, FM's, TM's, and TR's concerning training) .................... FM 21-6

e. List of Training Films, Film Strips, and Film Bulletins (listing TF's, FS's, and FB's by serial number and subject) .................... FM 21-7

f. Military Training Aids (listing Graphic Training Aids, Models, Devices, and Displays) .................... FM 21-8

### 127. STANDARD NOMENCLATURE LISTS.

a. Ammunition.
  Ammunition, fixed and semifixed, all types, including subcaliber for pack, light and medium field artillery, including complete round data .................... SNL R-1
  Ammunition for antiaircraft artillery .................... SNL P-5
  Ammunition instruction material for antiaircraft, harbor defense, heavy field, and railway artillery, including complete round data .................... SNL P-8

b. Care and Maintenance.
  Cleaning, preserving and lubricating materials; recoil fluids, special oils, and miscellaneous related items .................... SNL K-1
  Soldering, brazing and welding material, gases and related items .................... SNL K-2

## 40-MM AUTOMATIC GUN M1 (AA) AND 40-MM ANTIAIRCRAFT GUN CARRIAGES M2 AND M2A1

c. **Materiel.**

| | |
|---|---|
| Director, A.A., M5A2 and M6 (British) | SNL F-209 |
| Gun, automatic, 40-mm, M1 and carriage, gun 40-mm, M2 (A.A.) | SNL A-50 |
| Indicator, range, Mk. I, Navy | SNL F-287 |
| Quadrant, gunner's, M1 | SNL F-140 |
| Sight, computing, M7 | SNL F-276 |
| Sight, correctional, Mk. V | SNL F-286 |
| Small arms, automatic gun, trench mortar and field artillery sighting equipment and fire control instruments | SNL F-1 |
| System, remote control, M10 | SNL F-208 |
| Telescope, elbow, M17 (for directors M4, M5, M5A1, M6, M7, and M7A1 | SNL F-231 |
| Unit, generating, M5 and M6 | SNL F-227 |

## 128. EXPLANATORY PUBLICATIONS.

a. **Ammunition.**

| | |
|---|---|
| Ammunition, general | TM 9-1900 |
| Artillery ammunition | OS 9-20 |
| Range regulations for firing ammunition for training and target practice | AR 750-10 |

b. **Care and Preservation.**

| | |
|---|---|
| Artillery lubrication, general | OFSB 6-4 |
| Chemical decontamination materials and equipment | TM 3-220 |
| Cleaning, preserving, lubricating, and welding materials and similar items issued by the Ordnance Department | TM 9-850 |
| Cold weather lubrication and service of artillery materiel | OFSB 6-5 |
| Defense against chemical attack | FM 21-40 |
| Maintenance and care of pneumatic tires and rubber treads | TM 31-200 |
| Military chemistry and chemical agents | TM 3-215 |

c. **Fire Control Materiel.**

| | |
|---|---|
| Auxiliary fire control instruments | TM 9-575 |
| Directors M5A1, M5, and M6 | TM 9-659 |
| Fire control and position finding | FM 4-15 |
| Generating units M5 and M6 | TM 9-616 |
| Service of the piece, 40-mm fire unit | FM 4-160 |

d. **Firing tables** ............................................. FT 40-AA-A-2

**REFERENCES**

e. **Inspection and Maintenance.**

| | |
|---|---|
| AA cable system and repair kits, all types, and voltage controller M1 | TM 9-649 |
| Instruction guide: Instrument repairman | TM 9-2602 |
| Ordnance maintenance: Directors M5 and M6 | TM 9-1659 |
| Ordnance maintenance: Generating unit M5 | TM 9-1616 |
| Ordnance maintenance: Gun and top carriage of 40-mm antiaircraft gun materiel M2 | TM 9-1252 |
| Ordnance maintenance: Inspection | TM 9-1100 |
| Ordnance maintenance: Lower carriage of 40-mm antiaircraft gun materiel M2 | TM 9-1253 |
| Ordnance maintenance: Remote control systems M1 and M5 | TM 9-1643 |
| Ordnance maintenance: Telescopes and directors | TM 9-1652 |

# TM 9-252

## 40-MM AUTOMATIC GUN M1 (AA) AND 40-MM ANTIAIRCRAFT GUN CARRIAGES M2 AND M2A1

# INDEX

## A

| | Page |
|---|---|
| Accessories, gun and carriage | |
| care and preservation | 137 |
| description | 266 |
| Accidents (ammunition), field | |
| report of | 265 |
| Adjustment | |
| brakes | 150 |
| computing sight | 197, 201 |
| equilibrator | 145 |
| for single fire | 148 |
| rear firing pedal | 147 |
| recoil cylinder control rod valve | 108 |
| remote control system | 224 |
| weapon | 156 |
| wheel bearing | 185 |
| Ammeter (for electric brakes) | 266 |
| Ammunition | |
| authorized rounds | 255 |
| care, handling, and preservation | 254 |
| care in tropical climates | 284 |
| data | 5, 8, 12, 258, 260 |
| description | 260 |
| firing | 260 |
| fuzes | 264 |
| identification, and firing tables | 252 |
| loading in clips | 112 |
| lot number | 253 |
| marking | 253 |
| packing data | 264 |
| painting | 254 |
| Artillery gun book | 266 |
| Assembly | |
| breech ring closing spring | 176 |
| breechblock and breech ring | 173 |
| computing sights | 199 |
| gun (field strip) | 179 |
| equilibrator | 180 |
| wheels, hub, and brake drum | 182 |
| Axles | |
| description and functioning | 85, 92 |
| malfunctions and corrections | 127 |
| Azimuth | |
| bearings, lubrication | 134 |
| indicator | |
| description | 217 |

| | Page |
|---|---|
| differences among carriage | |
| models | 12 |
| lubrication | 251 |
| oil gears | |
| conversion | 244 |
| lubrication | 134 |
| switches | |
| adjustment | 232 |
| description | 217 |

## B

| | Page |
|---|---|
| Barrel assembly | |
| care and preservation | 139 |
| cleaning | 138 |
| data | 12 |
| description and functioning | 17 |
| carrier | 267 |
| inspection | 157 |
| installation | 161, 180 |
| modifications | 10 |
| precautions | 15 |
| removal | 161, 169 |
| replacement of abutment | 179 |
| Battery, dry-cell | |
| care and preservation | 149 |
| in low temperatures | 289 |
| Bell housing (remote control | |
| system), precautions | 16 |
| Blocking (rail shipment) | 282 |
| Bore, gun | |
| data | 12 |
| low temperature care | 287 |
| Bore sight, description and | |
| operation | 205 |
| Brakes | |
| adjustment | 150 |
| ammeter for electric brakes | 266 |
| care and preservation (electric | |
| brakes) | 149 |
| in low temperatures | 288 |
| control set M2 (electric brakes) | 268 |
| description and functioning | |
| electric brakes | 93 |
| hand brakes | 96 |
| inspection | 160 |

294

# INDEX

## B—Cont'd

| | Page |
|---|---|
| Brakes—Cont'd | |
| installation of drum | 183 |
| malfunctions and corrections (electric brakes) | 127 |
| operation | 119 |
| rail shipment | 280 |
| removal of brake drum | 182 |
| Breech mechanism | |
| care and preservation | 140 |
| cleaning | 138 |
| closing the breech | 99 |
| cooling slots | 11 |
| description and functioning | 30 |
| casing | 20 |
| side cover and lever | 34 |
| firing mechanism, low-temperature care | 288 |
| installation | 180 |
| lubrication | 134 |
| opening the breech | 98 |
| operation | 98 |
| automatic | 36 |
| manual | 35 |
| removal | 169 |
| storage and shipment (cover) | 279 |
| top cover damaged or broken | 125 |
| Breech ring | |
| assembly and disassembly | 173 |
| closing spring | 176 |
| description and functioning | 23 |
| outer crank assembly | 27 |
| safety plunger | 30 |
| installation | 179 |
| barrel catch | 176 |
| removal | 172 |
| barrel catch | 175 |
| replacement of parts | 178 |
| Breechblock | |
| assembly and disassembly | 173 |
| data | 12 |
| description and functioning | 26 |
| locking pin | 22 |
| failure to close | 121 |
| inspection | 158 |
| low-temperature care | 288 |
| modifications | 10 |
| operation | 35, 36 |
| removal and installation | 164 |

## C

| | Page |
|---|---|
| Cable connections, remote control system | 215 |
| Canvas, care of | 151 |
| Care and preservation | |
| ammunition (tropical climates) | 284 |
| carriage | 144 |
| cleaners and preserving materials | 151 |
| computing sights | 204 |
| general instructions | 136 |
| miscellaneous materials and tools | 152 |
| paints and related materials | 151 |
| tropical climates | 284 |
| washing | 152 |
| Carriages M2 and M2A1 | |
| accessories | 266 |
| care and preservation | 144 |
| characteristics | 8 |
| data | 13 |
| description and functioning | |
| axles and suspension | 85 |
| brakes | 93 |
| chassis frame | 80 |
| elevating mechanism | 74 |
| equilibrators | 76 |
| firing mechanism | 79 |
| hubs, wheels, and tires | 96 |
| lighting equipment | 97 |
| oil gear units | 79 |
| platform frame assembly | 74 |
| power synchronizing mechanism | 77 |
| top carriage | 73, 82 |
| traversing mechanism | 77 |
| differences among models | 11 |
| emplacement (low-temperature) | 288 |
| firing mechanism attached to | 36 |
| inspection | 158 |
| lubrication and cleaning | |
| in tropical climates | 283 |
| in low temperatures | 284 |
| operation under unusual conditions | 284 |
| precautions (firing position) | 14 |
| steering | 91 |
| storage and shipment | 277 |
| Cartridges | |
| authorized rounds | 256 |
| clip release arrangement | 51 |

# TM 9-252

## 40-MM AUTOMATIC GUN M1 (AA) AND 40-MM ANTIAIRCRAFT GUN CARRIAGES M2 AND M2A1

### C—Cont'd

| | Page |
|---|---|
| Cartridges—Cont'd | |
| description | |
| hand extractor | 274 |
| rammer assembly | 53 |
| remover | 268 |
| operation of rammer assembly | 53, 55 |
| partially ejected or not ejected case | 124 |
| removal of case | 117 |
| Characteristics | |
| carriage M2 | 8 |
| gun M1 | 5 |
| Chassis | |
| care and preservation | 146 |
| description | |
| frame | 80 |
| front chassis | 89 |
| rear chassis | 92 |
| lubrication | 135 |
| precautions in use of spring lock handles | 14 |
| Cleaners and preserving materials | 151 |
| Cleaning | |
| all parts | 131 |
| barrel exterior | 138 |
| bore (after firing) | 137 |
| breech mechanism | 138 |
| domestic shipment | 277 |
| in low temperatures | 285 |
| remote control system, precautions | 16 |
| Clips (carriage) (See Straps and clips (carriage)) | |
| Cocking lever, inner | |
| description | 37 |
| functioning | 38, 40 |
| installation, and removal | 173 |
| location of | 26 |
| Counterrecoil | |
| automatic firing cycle | 63 |
| control speed of | 115 |
| description and functioning | 72 |
| failure to or violent | 126 |
| slow | 125 |
| Crank assembly, breech ring | |
| description and functioning | 27 |
| functioning during firing | 40 |
| operation | 35, 36 |

| | Page |
|---|---|
| Crankshaft collar, breech ring, replacement | 178 |

### D

| | Page |
|---|---|
| Data | |
| ammunition | 258, 260 |
| packing data | 264 |
| carriages M2 and M2A1 | 13 |
| fuzes | 264 |
| gun M1 | 12 |
| Description | |
| ammunition | 260 |
| automatic firing cycle | 60 |
| automatic loading | |
| control mechanism | 55 |
| tray | 52 |
| axles and suspension | 85 |
| bore sight | 205 |
| brakes | 93 |
| breech casing | 20 |
| breech operating mechanism | 30 |
| breech ring | 23 |
| barrel catch | 25 |
| outer crank assembly | 27 |
| safety plunger | 30 |
| breechblock assembly | 26 |
| chassis frame | 80 |
| computing sights | 192 |
| elevating mechanism | 74 |
| equilibrators | 76 |
| extractor assembly | 30 |
| firing mechanism | 36, 37, 79 |
| fuzes | 264 |
| gun M1 | 17 |
| hubs, wheels, and tires | 96 |
| lighting equipment | 97 |
| oil gear units | 79 |
| platform and frame assembly | 74 |
| recoil mechanism | 70 |
| telescopes M7 and M74 | 195 |
| top carriage | 73 |
| traversing and power synchronizing mechanisms | 77 |
| Director dials, trouble shooting | 234 |
| Disassembly | |
| breech ring closing spring | 176 |
| breechblock and breech ring | 173 |
| computing sights | 199 |

# INDEX

## D—Cont'd

| | Page |
|---|---|
| Disassembly—Cont'd | |
| equilibrator | 180 |
| wheel, hub, and brake drum | 182 |
| Drawbar | |
| description and functioning | 90 |
| inspection | 160 |
| rail shipment | 280 |
| Dummy socket, differences among carriage models | 11 |

## E

| | Page |
|---|---|
| Ejector, shell | 275 |
| Elevating mechanism | |
| adjustment of limit stops | 231 |
| care and preservation | |
| elevating arc (in low temperatures) | 287 |
| elevating rack and pinion | 146 |
| description and functioning | 74 |
| elevation clutch lever | 216 |
| elevation limit switch | 215 |
| elevation of gun | 8 |
| by hand | 109 |
| by power | 111 |
| inspection | 159 |
| lubrication of gear cases | 135 |
| malfunctions and corrections | 126 |
| Equilibrator | |
| adjustment | 145 |
| assembly and disassembly | 180 |
| description and functioning | 76, 270 |
| lubrication | 134 |
| Extractor assembly | |
| description and functioning | 30 |
| hand extractor | 274 |
| spindle | 270 |
| installation | 167, 174 |
| removal | 166, 173 |

## F

| | Page |
|---|---|
| Feed control | |
| check and release mechanism | 57 |
| thumb lever to left (or right) | 122, 123 |
| Feed pawls (See Pawls; feed and stop pawls) | |
| Feed rods, description and functioning | 49 |

| | Page |
|---|---|
| Feed rollers, description and functioning of rollers and catch and plunger mechanisms | 50 |
| Fire control (See Sighting and fire control equipment) | |
| Firing | |
| cleaning the bore after | 137 |
| description and functioning (automatic firing cycle) | |
| counterrecoil | 63 |
| recoil | 60 |
| failure to single fire | 124 |
| functioning | 40 |
| observations during | 114 |
| operation prior to | 111 |
| placing gun in position for | 99 |
| positions of lever | |
| automatic fire | 43 |
| safe and single fire | 42 |
| precautions | 13 |
| preparation for | 260 |
| to fire the gun | 40, 112 |
| to place in position for | 99 |
| (See also Recoil and Counterrecoil) | |
| Firing mechanism | |
| care and preservation | 147 |
| data | 12 |
| description and functioning | 36, 79 |
| check and release mechanism | 40, 59 |
| firing lever and pawl | 41 |
| parts mounted on breech casing | 41 |
| rammer catch mechanism | 40 |
| lubrication | 134 |
| parts attached to carriage | 36 |
| Firing pin | |
| data | 13 |
| description | 37 |
| functioning | 40 |
| installation and removal | 173 |
| low-temperature care | 288 |
| Firing tables | 252 |
| Flash hider | |
| description | 17 |
| wrench | 271 |
| Fuzes, description and data | 264 |

TM 9-252

**40-MM AUTOMATIC GUN M1 (AA) AND 40-MM ANTIAIRCRAFT GUN CARRIAGES M2 AND M2A1**

## G

| | Page |
|---|---|
| Gun M1 | |
| accessories | 266 |
| care and preservation | 139 |
| characteristics | 5 |
| data | 12 |
| description and functioning | 17 |
| gun stay | 92 |
| junction box | 215 |
| striker protrusion gage | 277 |
| develops low top speed | 234 |
| elevating | 109 |
| field strip assembly | 179 |
| field stripping | 168 |
| inspection | 157 |
| gun stay | 160 |
| lubrication and cleaning | |
| low-temperature | 284 |
| tropical climates | 283 |
| modifications | 10 |
| operation | 98 |
| placing in firing position | 99 |
| placing in traveling position | 117 |
| prior to firing | 111 |
| under unusual conditions | 284 |
| register mark on gun tube | 11 |
| rough or jerky in tracking | 235 |
| serial number | 156 |
| storage and shipment | 277 |
| to unload | 116 |
| traversing | 108 |
| (See also Cocking; Firing; Loading mechanism; Recoil; and Counterrecoil) | |

## H

| | |
|---|---|
| Hubs | |
| description and functioning | 96 |
| installation | 183 |
| removal | 182 |
| Hydraulic oil gears, lubrication | 134 |

## I

| | |
|---|---|
| Inspection | |
| carriage | 158 |
| gun | 156 |
| purpose of | 157 |
| serial numbers | 156 |
| storage and shipment | 279 |
| visual inspection upon receipt | 156 |

## J

| | Page |
|---|---|
| Jacks, leveling | |
| description and functioning | 84 |
| operation | 118 |

## L

| | |
|---|---|
| Level assemblies, care and preservation | 148 |
| Lighting equipment | |
| care and preservation | 150 |
| description and functioning | 97 |
| devices | 204 |
| inspection of lights | 160 |
| malfunctions and corrections | 129 |
| (See also Stop lights and Taillights) | |
| Loader, automatic | |
| description and functioning | |
| control mechanism | 55 |
| feed mechanism | 44, 274 |
| hood and shield | 267 |
| lifters | 271 |
| loader | 17, 22, 44 |
| loading tray | 52 |
| feed rollers fail to rotate | 125 |
| inspection | 158 |
| installation | 179 |
| lubrication | 134 |
| precautions | 15 |
| removal | 169 |
| (See also Loading) | |
| Loading | 112 |
| for rail shipment | 279 |
| malfunctions and corrections | 122, 123 |
| precautions | 13 |
| (See also Loader, automatic) | |
| Lubrication | |
| azimuth indicator | 251 |
| computing sights | 204 |
| equipment | 131 |
| for domestic shipment | 277 |
| guide | 130 |
| in tropical climates | 283 |
| low-temperature | 284 |
| oil can points | 135 |
| points of application | 131 |
| precautions | 14 |
| remote control system | 247 |
| reports and records | 136 |
| supplies | 130 |

# INDEX

## M

| | Page |
|---|---|
| Magnet, facing of | 149 |
| Malfunctions and corrections | |
|   axles | 124 |
|   breechblock | 121 |
|   elevating and traversing mechanisms | 126 |
|   feed control thumb lever | 122, 123 |
|   gun | |
|     fails to fire | 122, 123 |
|     misfires | 120 |
|   lights | 130 |
|   recoil mechanism | 125 |
|   remote control system | 232 |
|   wheels and tires | 127 |
| Misfire (gun) | 120 |
| Muzzle covers, storage and shipment | 279 |

## O

| | Page |
|---|---|
| Oil can points | 136 |
| Oil gear | |
|   conversion | 244 |
|   description and functioning | 79, 215 |
|   low-temperature care | 288 |
|   malfunctions and corrections | 232, 233 |
|   removal of units | 236 |
|   replacement of units | 243 |
| Operation | |
|   arctic climates | 284 |
|   bore sight | 205 |
|   brakes | 119 |
|   breech mechanism | 35, 36, 98 |
|   cartridge rammer rod assembly | 53, 55 |
|   computing sights | 197 |
|   moist or salty atmosphere or sand or dust conditions | 289 |
|   remote control system M5 | 219 |
|   tropical climates | 283 |
| Outriggers | |
|   description and functioning | 83 |
|   differences among carriage models | 11 |
|   rail shipment (covers) | 282 |

## P

| | Page |
|---|---|
| Painting | |
|   ammunition | 254 |
|   lubrication devices | 155 |
|   paint and related materials | 151 |
|   preparation for painting | 153 |
|   removal of paint | 155 |
|   storage and shipment | 279 |
| Pawls, description and functioning | |
|   catch release pawl | 51 |
|   feed and stop pawls | 48 |
|   firing lever pawl | 41 |
| Percussion mechanism | |
|   description | 37 |
|   functioning | 38 |
| Placarding material | 280 |
| Platform and frame assembly, description and functioning | 74 |
| Plungers | |
|   description and functioning | |
|     breech ring safety plunger | 30 |
|     key puller | 270 |
|     check plunger | 39, 40 |
|     feed roller plunger mechanism | 51 |
|   installation and removal (check plunger) | 173 |
|   replacement (breech ring safety plunger) | 178 |
| Power shaft adapter gears and bearings, lubrication | 134, 135 |
| Power synchronizing mechanism | 77 |

## R

| | Page |
|---|---|
| Rammer catch and check mechanisms | 40 |
| Rammer cocking lever shaft assembly, description and functioning | 57 |
| Rammer rod, cartridge, description and operation | 52, 53 |
| Rammer shoe, malfunctions and corrections | 122, 123, 124 |
| Ramps (loading for rail shipment) | 279 |
| Range | |
|   data | 13 |
|   maximum effective | 5 |
| Recoil | |
|   automatic firing cycle | 60 |
|   data | 13 |
|   description and functioning | 70 |
|   length of | 141 |
|   measuring length of | 114 |
|   slow, violent, or excessive in length | 125 |

TM 9-252

**40-MM AUTOMATIC GUN M1 (AA) AND 40-MM ANTIAIRCRAFT GUN CARRIAGES M2 AND M2A1**

### R—Cont'd

| | Page |
|---|---|
| Recoil cylinder | |
| check liquid in | 108 |
| emptying and exercising | 142 |
| filling | 141 |
| inspection | 158 |
| installation | 180 |
| low-temperature care | 288 |
| modifications (fluid) | 10 |
| removal | 168 |
| Recoil mechanism | |
| care and preservation | 140 |
| care of oil | 143 |
| description and functioning | |
| counterrecoil | 72 |
| recoil | 70 |
| lubrication | 134 |
| malfunctions and corrections | 125 |
| Recuperator spring, description and functioning | 17 |
| Remote control system M5 | |
| azimuth switch | 232 |
| changing oil and renewing filter bobbin assembly | 249 |
| description | 206 |
| elevation limit stops | 231 |
| lubrication | 247 |
| maintenance precautions | 16 |
| malfunctions and corrections | 232 |
| operation | |
| faults and precautions | 16, 223 |
| setting up and orienting | 219 |
| tests and adjustments | |
| accuracy test, dither adjustment, and backlash | 225 |
| neutral adjustment | 227 |
| resetter spiral gear | 228 |
| torque test | 228 |
| Reports and records, ammunition accidents | 265 |
| Rifling data | 12 |

### S

| | Page |
|---|---|
| Safety lever, outer, functioning | 42 |
| Safety switch | |
| description and functioning | 95 |
| differences among carriage models | 11 |
| Serial numbers (gun and tube) | 156 |

| | Page |
|---|---|
| Shell, description of | |
| ejector | 274 |
| pusher | 276 |
| Shipment (See Storage and shipment) | |
| Sighting and fire control equipment | |
| arrangement of | 187 |
| direct fire sights | 189, 191 |
| low-temperature care | 288 |
| Sighting system M3, carriages equipped with | 12 |
| Sights M7 or M7A1, computing | |
| care, preservation, and lubrication | 204 |
| carriages equipped with M7 | 12 |
| direct fire sights | 189, 191 |
| description | 192 |
| disassembly and assembly | 199 |
| operation and tracking adjustments | 197 |
| lighting devices | 204 |
| precautions | 15 |
| tests and adjustments | 201 |
| Spare parts, organizational | 137, 265 |
| Spring, closing, removal and installation | 164 |
| Steering (carriage) | 91 |
| Stop light | |
| care and preservation | 150 |
| malfunctions and corrections | 128 |
| replacement of lamps | 150 |
| Storage and shipment | |
| domestic shipment | 277 |
| rail shipment | 279 |
| limited storage | 282 |
| Straps, safety, precautions | 14 |
| Straps and clips (carriage), differences among carriage models | 11 |
| Suspension, description and functioning | 85, 88 |

### T

| | Page |
|---|---|
| Taillights, replacement of lamps | 150 |
| Targets, tracking | |
| changing | 223 |
| tracking | 222 |
| Telescopes M7 and M74 | |
| care and preservation | 205 |
| description | 195 |

## INDEX

### T—Cont'd

Tires
- care and preservation ... 148
  - low temperatures ... 289
  - tropical climates ... 283
- data ... 13
- description and functioning ... 96
- differences among carriage models ... 12
- inspection ... 160
- malfunctions and corrections ... 127
- precautions ... 15
- rail shipment ... 280
- removal and installation ... 185

Tools ... 152
Torque test, remote control system ... 228
Tracking
- gun rough or jerky ... 235
- practice ... 224
- the target ... 222

Traversing mechanism
- description and functioning ... 8, 77
- lubrication of bearings and racks ... 134, 135
- malfunctions and corrections ... 126
- traversing by
  - hand ... 108
  - power ... 109

Tubes, gun
- care and preservation ... 148
- register mark ... 17
- modifications ... 11

### U

Unloading the gun ... 116

### V

Valve, pilot
- removal ... 235
- replacement ... 237

### W

Wheels
- adjustment of bearing ... 185
- care and preservation ... 150
- description and functioning ... 96
- differences among carriage models ... 12
- inspection ... 160
- installation ... 185
- lubrication of bearings ... 134
- malfunctions and corrections ... 127
- removal ... 182

Wiring connections, malfunctions and corrections ... 232

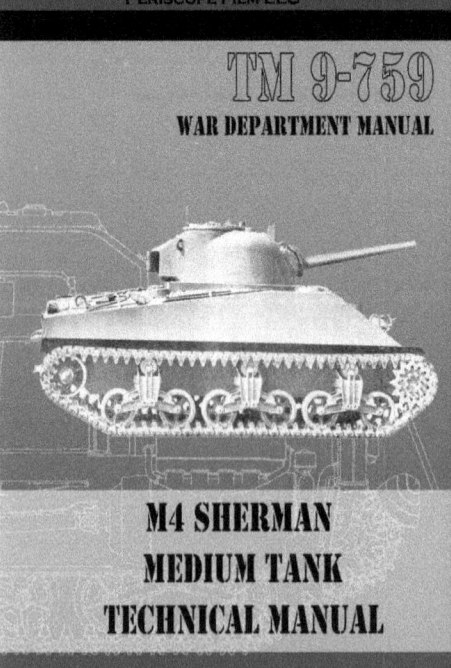

## Also Now Available!

**Visit us at:**

**www.PeriscopeFilm.com**

©2013 Periscope Film LLC
All Rights Reserved
ISBN#978-1-937684-41-9
www.PeriscopeFilm.com

www.ingramcontent.com/pod-product-compliance
Lightning Source LLC
Chambersburg PA
CBHW070531160426
43199CB00014B/2244